Get Qualified: Inspection and Testing

The *Get Qualified* series provides clear and concise guidance for people looking to work within the electrical industry. This book outlines why the inspection and testing of electrical installations is important, and what qualifications are required in order to test, inspect and certify. All you need to know about the subject of inspection is covered in detail, making this book the ideal guide for those who are new to the subject and experienced professionals alike. There are also sections on exam preparation, revision exercises and sample questions.

Kevin Smith is a highly experienced Electrical Trainer, who in 2006 started his own electrical training company, which led him to deliver training courses for companies large and small throughout the UK. Since 2008 he has personally taught the 17th Edition *IET Wiring Regulations* (BS 7671) to over 1000 electricians. Kevin is currently the Training Manager for Seaward, a large UK-based manufacturer of electrical test equipment. He is the author of *Get Qualified: Portable Appliance Testing* also published with Routledge.

Get Qualified:
Inspection and Testing

Kevin Smith

Routledge
Taylor & Francis Group

LONDON AND NEW YORK

First published 2018
by Routledge
2 Park Square, Milton Park, Abingdon, Oxon OX14 4RN

and by Routledge
711 Third Avenue, New York, NY 10017

Routledge is an imprint of the Taylor & Francis Group, an informa business

British Library Cataloguing-in-Publication Data
A catalogue record for this book is available from the British Library

Library of Congress Cataloging in Publication Data
A catalog record has been set up for this book.

ISBN: 978-1-138-31026-1 (hbk)
ISBN: 978-1-138-18963-8 (pbk)
ISBN: 978-1-315-56108-0 (ebk)

Typeset in Kuenstler
by Servis Filmsetting Ltd, Stockport, Cheshire

Contents

PART 1

CHAPTER 1 Introduction 3

CHAPTER 2 What is inspection and testing? 7

CHAPTER 3 Who can carry out inspection and testing? 11

CHAPTER 4 What are my training options? 17

PART 2

CHAPTER 5 Inspection and testing safety 27

CHAPTER 6 Safe isolation 31

CHAPTER 7 Information required for inspection and testing 41

CHAPTER 8 The difference between initial verification and periodic
inspection and testing 45

CHAPTER 9 Inspecting an electrical installation 55

CHAPTER 10 Testing an electrical installation 63

CHAPTER 11 Testing the continuity of conductors (regulation 612.2) 71

CHAPTER 12 Testing insulation resistance (regulation 612.3) 89

CHAPTER 13 Testing polarity (regulation 612.6) 99

CHAPTER 14 Testing earth electrode resistance (regulation 612.7) 103

CHAPTER 15 Protection by automatic disconnection
of supply (regulation 612.8) 107

CHAPTER 16 Testing earth fault loop impedance (regulation 612.9) 113

CHAPTER 17 Testing additional protection (regulation 612.10) 121

CHAPTER 18 Testing prospective fault current (regulation 612.11)...................... 125

CHAPTER 19 Check of phase sequence (regulation 612.12).............................. 129

CHAPTER 20 Functional testing (regulation 612.13) 131

CHAPTER 21 Verification of voltage drop (regulation 612.14) 133

CHAPTER 22 Inspection and testing certification 137

PART 3

CHAPTER 23 Good examination practice and revision... 143

CHAPTER 24 Sample examination questions 147

INDEX 163

Part 1

Part 1

CHAPTER 1

Introduction

It is a fundamental principle of the *IET Wiring Regulations* (BS 7671) that all electrical installations must meet two basic criteria: first, they must be safe and, second, they must perform their intended function, i.e. they must work. In the early days of electrical installations this was simple to achieve, because electrical installations were very simple in nature and safety requirements very basic by today's standards.

Later, as the electrical industry grew, most electrical installation work was carried out by local electricity boards using time-served tradesmen, who were not hurried by the time and cost pressures experienced by

Get Qualified: Inspection and Testing. 978 1 138 18963 8 © Kevin Smith. Published by Taylor & Francis.

modern electrical contractors. As a result, most installation work was performed to a high standard and therefore the checks carried out on this work were minimal.

Fast forward to today and we see modern, highly complicated electrical installations, installed very quickly using teams of installers, many of whom will not be time-served electricians. Add to this the increasing demands placed upon us to improve health and safety and the growing culture of claims and litigation. It is therefore essential in the modern world that the safety and correct functioning of electrical installations is not just taken for granted, but is thoroughly proven by a rigorous process of inspection and testing. Out of this basic need has grown the art of inspection and testing, the subject on which this book is based.

Inspection and testing has now become an integral part of the work carried out by all electricians. A test meter and a pad of certificates are now as important to an electrician's daily work as are his or her screwdrivers, wire cutters and SDS drill. While inspection and testing has long been a requirement of the *IET Wiring Regulations* (BS 7671), widespread adoption of inspection and testing throughout the industry has only occurred due to the hard work of trade organisations such as NICEIC and others. The addition of electrical safety into the *Building Regulations (Part P)* in 2005 made a huge difference to the amount of inspection and testing performed within domestic installations, which had long been a part of the industry dominated by DIY and cowboy electricians.

There are, however, sectors of the industry still slow to fully embrace inspection and testing; from my experience one notable example being maintenance electricians in production or factory environments, where it is common to see electrical repairs carried out without any inspection, testing or certification. It is key to improving safety within the electrical industry that the practices of inspection and testing are adopted by all sectors across the industry and that the importance of inspection and testing is understood by all.

In this book I aim to simplify this subject and introduce you to each topic in a straightforward and common-sense way. I will look at what is required to meet your legal obligations and explain the options you have to 'Get Qualified' and undertake inspection and testing in your own right. Throughout the book I will make frequent reference to statutory documents, like the *Electricity at Work Regulations 1989* and to other guidance documents, such as the *IET Wiring Regulations* (BS 7671) and *IET Guidance Note 3*. A list of these documents can be found in Chapter 7, 'Information required for inspection and testing'. Many documents are available as free downloads and I would strongly suggest you use them as an accompaniment while you read the rest of this book.

The ability to undertake inspection and testing is not rocket science and is easily within the reach of most electricians. With the right training and experience you will very soon be performing this task quickly and efficiently. You will also have the added peace of mind that this important safety task has been done correctly and in accordance with current industry best practice.

CHAPTER 2

What is inspection and testing?

It may appear obvious to say that the subject of this book 'inspection and testing' is made up of two activities, 'inspection' and 'testing'. Inspection involves using your senses – sight, touch, hearing, etc. – to inspect an electrical installation. Testing is the process of using a meter to take measurements. Always be mindful, when discussing the subject of 'inspection and testing', to consider these two activities separately as they both contribute in different ways to our knowledge of the electrical installation.

The aim of the inspection and testing process is to establish the safety and correct functioning of the electrical installation, without posing a danger to the inspector or other persons and without damaging the electrical installation under test.

To meet this basic and essential aim, the inspection and testing process for new installations usually follows the format detailed below:

1. Safe isolation
2. Inspection   } Electrical supply safely disconnected
3. Dead testing

4. Live testing   } Electrical supply connected
5. Functional testing

The aim of the above process is to gradually, step-by-step, prove the safety of the electrical installation in the safest way possible. Only when the installation successfully passes one stage within the process, is it then safe to proceed to the next step. It is essential that inspection and testing are carried out throughout the installation process and on completion. It is not possible to carry out full inspection and testing on a completed installation, as much of the installation may be hidden within the fabric of the building and we may not be able to carry out a full range of tests with all circuit loads connected. Many electricians make the mistake of believing that inspection and testing is something done at the end of a job, as a type of final sign-off procedure. It is not possible, for the reasons stated above, to fully inspect and test completed installations, and attempting to perform inspection and testing in this way is both incorrect and inefficient.

TYPES OF INSPECTION AND TESTING

There are two main types of inspection and testing referred to in the *IET Wiring Regulations* (BS 7671). The first is **initial verification**, the inspection and testing of new installations, alterations or additions. The second is **periodic inspection**, performed on existing installations.

Initial verification

A new electrical installation, prior to initial verification, has never been connected to its supply or performed its intended function before. It is possible therefore that this new installation may contain defects introduced during the installation process, which must be identified and corrected before the installation is put into service. All new installations must comply with the current edition of the *IET Wiring Regulations* (BS 7671); this compliance must also be confirmed at this stage. The initial verification of new electrical installations is essential, as any issues not detected at this stage are likely to remain undetected for many years and could pose a hidden danger to the safety of those using the installation in the future. The initial verification process is also applied to additions and alterations to an existing electrical installation. These additions or alterations may include, for example, the addition of a new shower circuit, the alteration of an existing socket circuit to include more socket outlets or even the repair or replacement of a faulty luminaire. In each example the purpose of initial verification is the same: to prove the safety of the new installation or equipment and ensure compliance with the regulations.

Periodic inspection

As electrical installations age, some damage or deterioration is inevitable. The requirements of the *IET Wiring Regulations* are also being constantly updated, making many older installations appear obsolete by modern standards. The purpose of a periodic inspection is to ensure that an existing electrical installation is safe for continued use. By its very nature a periodic inspection is much more difficult and nuanced to carry out than the very black-and-white initial verification process. Electrical installations receiving a periodic inspection will not be in 'as new' condition and are unlikely to fully comply with the latest edition of the wiring regulations. A greater degree of judgement is therefore required from the inspector, who must consider additional factors such as context, rates of deterioration and other evidence before making final recommendations to the customer.

CHAPTER 3

Who can carry out inspection and testing?

This question is often the subject of much confusion, so in this chapter I will attempt to simplify the requirements and help you to finally get an answer to the question above.

Unfortunately, the confusion around this subject is easy to understand. There is not one single, easy criteria that we must meet in order to be allowed to carry out inspection and testing. Instead, there is a myriad of different requirements imposed by many different organisations and only by interpreting these requirements as a whole, and applying them to the sector of the electrical industry in which we work, can we finally shed some light on this question.

Remember that inspection and testing is not a job for the inexperienced or untrained.

SKILLED PERSON

Recently, with the introduction of Amendment 3 of the *IET Wiring Regulations* (BS 7671:2008) in 2015, we have seen a change in the terminology used to relate to those carrying out inspection and testing. We now use the term 'skilled person'. Although this term has been in the regulations since 2008, its definition was weakened to allow room for the more familiar term 'competent person' to be used. The term 'competent person' is more familiar in the UK, as it aligns with the same term used in the *Electricity at Work Regulations 1989*. 'Competent person' has now been deleted from the wiring regulations and the 'skilled person' definition has been reinforced to take its place.

The current definition taken from BS 7671:2008 is:

> **Skilled person (electrically).** *Person who possesses, as appropriate to the nature of the electrical work to be undertaken, adequate education, training and practical skills, and who is able to perceive risks and avoid hazards which electricity can create.*

It can be seen from the definition above that a skilled person requires education, training and practical skills, in addition to the safety requirements, to perceive risk and avoid electrical hazards. The definition also refers to the nature of electrical work undertaken. It is reasonable to assume that where the nature of electrical work relates to inspection and testing the above requirements apply in this area.

The two regulations below, again taken from BS 7671:2008, detail the requirements for those carrying out initial verification and periodic inspection and testing.

> **610.5** – *The verification shall be made by a skilled person, or persons, competent in such work.*

> **621.5** – *Periodic inspection and testing shall be undertaken by a skilled person, or persons, competent in such work.*

You can see from the above that the term 'skilled person' is now used, although the word 'competent' is still included to aid crossover with the *Electricity at Work Regulations 1989*. Therefore to comply with the *IET Wiring Regulations* (BS 7671) any person wishing to carry out inspection and testing must meet the definition of a 'skilled person' in relation to the activity of inspection and testing.

COMPETENT PERSON

Although this term has now been removed from the *IET Wiring Regulations* (BS 7671), it is still an important part of electrical law in Great Britain. Almost all people wishing to undertake inspection and testing in the UK will be subject to the legal requirements contained in the *Electricity at Work Regulations 1989*, which applies to employers, employees and the self-employed.

Regulation 16 of the *Electricity at Work Regulations 1989* (below) is often referred to when attempting to quantify competence.

> *No person shall be engaged in any work activity where technical knowledge or experience is necessary to prevent danger or, where appropriate, injury unless he/she possesses such knowledge or experience, or is under such degree of supervision as may be appropriate having regard to the nature of the work.*

So the judgement is: has this person got enough knowledge and experience to avoid danger and injury, to themselves and others? And, just as importantly, could you prove it in court if it all goes wrong?

Establishing a person's competence is tricky and judgements must be made against two main criteria: **knowledge** and **experience**.

The *Electricity at Work Regulations 1989* are not specific to inspection and testing and lack detail. Knowledge of what? How much experience? Ultimately, these are decisions for the person responsible, often referred to as the 'Duty Holder', but we can go some way to making that judgement easier.

FIGURE 3.1 Competency is still an important part of electrical law in Great Britain

First, it is clear that there is no golden ticket that allows some to carry out inspection and testing and others not. Proving competence is about compiling evidence of relevant knowledge and experience; the more evidence we have the better able we are to call ourselves 'competent' and the happier an employer will be to allow us to undertake work on their behalf. Also remember that this is an ongoing process and as new guidance is published and new test equipment brought out, we must update our knowledge and experience to remain 'competent'.

Competent person schemes

Since 2005, in England and Wales, electrical safety in dwellings has been incorporated into the building regulations. (Similar provision also exists in Scotland and Northern Ireland.) *Approved Document P*, better known as Part P of the building regulations, lays out the standards for electrical installations in dwellings, and requires that some types of electrical work (listed below) carried out on these installations are notified to the local building control body.

Work in dwellings that requires notification in England:

- The installation of a new circuit.
- The replacement of a consumer unit.
- Any addition or alteration to existing circuits in a special location (bathroom, swimming pool or room containing a sauna heater).

Any work deemed 'notifiable' under the requirements of Part P must be certified as complying with the building regulations and notified to the local building control body. The owner of the dwelling will then receive a building regulations compliance certificate, which can be used in the future to prove compliance of the work.

Since Part P was updated in 2013, there are now three methods available for certification and notification of works within the scope of Part P.

1. Self-certification by a registered competent person.
2. Third-party certification by a registered third-party certifier.
3. Certification by a building control body.

All of the certification routes listed above still require that the electrical installation work complies with the requirements of BS 7671, which includes inspection and testing and the use of skilled persons.

Routes 2 and 3 allow for work to be checked and signed off by third parties, either a third-party certifier, usually an experienced electrical contractor who is a member of a third-party certification scheme, or a representative from your local building control body. In both these cases the installer is not required to be a member of a competent person scheme in their own right; they must still however comply with the requirements of BS 7671 and the building regulations.

Route 1 above, usually taken by professional electricians working in the domestic sector, gives the installer the ability to certify their own work as complying with the building regulations. In this route no third parties are required to certify the work; however to self-certify the installer must be a member of an approved competent persons scheme. Competent person schemes are available from a variety of companies, including ELECSA, NAPIT, NICEIC and Stroma. Those wishing to join a competent person scheme will be required to demonstrate competence to the scheme operator and comply with scheme rules. Scheme members are also required to pay a sizeable membership fee.

Those wishing to carry out regular work in domestic dwellings should pay close attention to scheme entry criteria when considering their inspection and testing training requirements. Most schemes will require you to hold or be working towards a formal inspection and testing qualification at Level 3.

CHAPTER 4

What are my training options?

Training is about building on the knowledge and experience you already have, therefore each person's training needs will be different. You must also consider the end goal – there is no point training to run a marathon if all you need to do is walk to the shops. Many people go for a course that is beyond their current ability level, subsequently falling at the first hurdle and spending a lot of money unnecessarily. It is always better to start small and build up as your knowledge and experience grow.

Essentially you will need to concentrate on the areas below:

- **Knowledge**
  - Electrical safety
  - Electrical theory
  - The *IET Wiring Regulations* (BS 7671)
  - *IET Guidance Note 3*.

- **Experience**
  - Practical electrical work
  - Using electrical test equipment
  - Inspecting electrical installations
  - Testing electrical installations
  - Completing electrical certification.

The above points form the basis of most inspection and testing courses and, depending on the level of each course, they will cover each topic in more or less detail. Some courses concentrate heavily on specific areas, such as initial verification or periodic inspection.

Unfortunately inspection and testing is not a subject that can be learned in isolation. Just as it is impossible to start building a wall 1m from the ground, it is impossible to learn inspection and testing without the solid foundations provided by knowledge of the wiring regulations and electrical theory, and practical experience within the electrical industry.

Most electricians will have received some form of inspection and testing training as part of their original apprenticeship or college-based training course. However, those entering the industry by other less traditional routes may be totally new to the subject. Whatever your route into the industry, or reasons for wanting to improve your inspection and testing skills, you must first ensure that you are starting the process from solid foundations.

All inspection and testing qualifications require candidates to have an excellent and up-to-date knowledge of the current *IET Wiring Regulations*; those not possessing such knowledge should strongly consider sitting a wiring regulations course before starting any inspection and testing training. In addition to knowledge of the wiring regulations you should also consider revising your knowledge of electrical theory, electrical health and safety requirements and, where necessary, your level of industry experience. Those who struggle with inspection and testing usually do so because of a lack of prior knowledge or experience. Do not rush into an inspection and testing course until you are happy you have done everything you can to prepare.

Whichever course option you decide to pursue, you will need to ensure that you have adequate free time for study, revision and practical practice. Where possible, the cooperation of your employer can also be a great help. Some employers will appoint a colleague who is experienced in inspection and testing to act as a mentor while you are on the course. Also letting your employer know the dates of your course and assessments in advance will ensure that there are no diary or workload problems during this important time.

INSPECTION AND TESTING QUALIFICATIONS

Formal inspection and testing qualifications are available from three main awarding organisations; these are **City & Guilds**, **EAL** and **LCL**. Training courses for these qualifications are delivered by approved training providers. While City & Guilds is the most recognisable of the three, each holds equal standing within the industry and there are only minor differences between them.

The content of these inspection and testing training courses is set nationally, with each organisation listing the specific criteria that must be achieved and the assessments that must be passed before a certificate will be issued.

Training providers wishing to offer these courses must first be approved to do so by the awarding organisation (City & Guilds, EAL or LCL) who will then carry out regular monitoring visits to ensure that the training provider operates to the high standards expected.

City & Guilds, **EAL** and **LCL** courses follow the same course structure. This structure is set nationally by Ofqual, ensuring parity between awarding organisations. These formal courses are also heavily based on the content of the *IET Wiring Regulations* (BS 7671) and *IET Guidance Note 3*, which are used extensively throughout these courses. If you intend to undertake these formal training courses you must have a copy of the latest version of the *IET Wiring Regulations* (BS 7671), *IET Guidance Note 3* or, in the case of the Level 2 qualifications, a copy of the *IET On-Site Guide*. All of these documents are also heavily referenced in this book and will be essential to your studies.

Background

The first, and probably best-known, stand-alone inspection testing qualification was the **City & Guilds 2391**. Introduced in the 1990s this qualification was designed to improve the standards of inspection and testing within the electrical industry. Written to cover both initial verification and periodic inspection, 2391 was accredited as a Level 3 qualification, similar in difficulty to an A-level. In the early days, 2391 courses were mainly attended by supervisors and managers. At the time it was felt that the course's detailed inspection and testing content was beyond that required by the average electrician. The City & Guilds 2391 qualification, usually delivered as a ten-week evening class, came with a formal practical assessment and a fearsome closed-book written exam. For the next decade the 2391 remained largely unchanged and became the benchmark for those performing inspection and testing within the electrical industry.

However, as time went on, typical 2391 candidates changed from being experienced electrical supervisors or managers to general electricians and, finally, to those with very limited electrical experience. To cope with this change in clientele, training providers increased the reliance on past papers and teaching in the test. This tactic triggered an arms race with the City & Guilds examiners, who, in an attempt to maintain standards, devised new and increasingly more difficult exam questions. Sadly, the result of this Cold War was that national pass rates for the 2391 written exam sank, on some occasions to below 40 per cent, earning the qualification a reputation for being 'un-passable' and scaring off many potential candidates.

Following the introduction of Part P into the building regulations in 2005, it was felt that there was a need for an inspection and testing qualification to support the growing numbers of Part P domestic electricians. Because of the 2391's fearsome reputation and the fact that it also covered periodic inspection and three-phase installations, not generally required in domestic installations, it was decided that the new qualification would be required.

Still available today, this new qualification, **City & Guilds 2392**, was to be positioned at Level 2, making it more accessible, and would cover only the initial verification of single-phase installations. The 2392 would be assessed by a multiple-choice on-line examination and a short practical assessment. In choosing an on-line exam for this new course, City & Guilds removed the need for training providers to stick to the fixed written exam timetables, also avoiding the long wait for exam results. Despite its obvious advantages the 2392 struggled to step outside of the shadow cast by its big brother, the 2391, never really achieving the popularity expected at its launch.

Today, I still strongly recommend the City & Guilds 2392 as a starting point for anyone with limited inspection and testing experience. As you will see later in this book, much of the content of the 2392 is directly transferable to the Level 3 inspection and testing qualifications and it is my experience that those candidates progressing to Level 3, having first successfully completed the Level 2 2392, find the course easier and achieve better exam results.

Taking advantage of the buzz around the launch of the 17th edition of the *IET Wiring Regulations* (BS 7671) in 2008, a new player emerged on the inspection and testing scene. With a long history and a strong brand in the engineering sector, EAL decided to launch their own range of electrical installation qualifications to directly compete with the established market leader, City & Guilds. These EAL courses mirrored the City & Guilds offering, with only minor differences in the exam and assessment methods. Later, LCL would also launch a range of Ofqual-accredited inspection and testing qualifications. In response to this new competition, City & Guilds finally decided that it was time to address the issues with the ageing 2391 qualification.

In an attempt to make the new 2391 qualification more manageable it was decided to divide the qualification in two; therefore replacing the 2391 with two new Level 3 qualifications, **City & Guilds 2394**, covering initial verification and the **City & Guilds 2395**, covering periodic inspection. Both new qualifications would share a common multiple-choice on-line exam, covering basic theory content. Each qualification would also have its own written exam and practical assessment. Initially, it took a while for these new qualifications to find their feet. Customers were confused about which qualification they needed to do and felt aggrieved that they now needed to pay for two courses instead of one. Also the new written exams were proving to be as difficult as the old 2391 exam and early pass rates were still very low. This was addressed in August 2014 when the written exam was extended to two hours, moving the pass rate closer to 60 per cent.

A further improvement was seen in 2016, with the addition of on-line short-answer examinations, introduced to replace the written exam, which is currently being phased out. The on-line short-answer exams finally offer candidates the chance to sit exams on demand, receive their results usually within a couple of weeks and, where necessary, take re-sits straightaway. Because of the computer-based short-answer format, it is my view that these exams are also easier than their written counterpart. For all the reasons listed above there has never been a better time to undertake a Level 3 inspection testing course.

In the table below I have attempted to summarise the main inspection and testing qualifications currently accredited by Ofqual. Full details of these qualifications can be found on the Ofqual website (https://www.gov.uk/government/organisations/ofqual) or on the websites of the individual awarding organisations.

Table 4.1 Qualifications currently available

Subject	Level	Awarding organisation	Qualification
Initial verification single phase	2	City & Guilds	Certificate in Fundamental Inspection, Testing and Initial Verification – 2392 (500/3516/2)
	2	EAL	Award in Fundamental Inspection, Testing and Initial Verification (603/0144/2)
Initial verification single phase and three phase	3	City & Guilds	Award in the Initial Verification and Certification of Electrical Installations – 2394 (600/3785/4)
	3	EAL	Award in the Initial Verification and Certification of Electrical Installations (600/4337/4)
	3	LCL	Award in the Initial Verification and Certification of Electrical Installations (601/5929/7)
Periodic inspection and testing	3	City & Guilds	Award in the Periodic Inspection, Testing and Certification of Electrical Installations – 2395 (600/4693/4)
	3	EAL	Award in the Periodic Inspection, Testing and Certification of Electrical Installations (600/4338/6)
	3	LCL	Award in the Periodic Inspection, Testing, Condition Reporting and Certification of Electrical Installations (601/5930/3)
Both initial verification and periodic inspection	3	LCL	Award in the Initial Verification, Periodic Inspection, Testing, Condition Reporting and Certification of Electrical Installations (601/5928/5)

CHOOSING A QUALIFICATION

From Table 4.1 you can see that inspection and testing qualifications break down into three types:

1. **Level 2 introductory** qualifications aimed at the domestic installer market or those who are new to the subject altogether.
2. **Level 3 initial verification** qualifications aimed at those wanting to inspect and test more complex new installations.
3. **Level 3 periodic inspection** qualifications aimed at those who will be completing Electrical Installation Condition Reports (EICRs) for existing installations.

While there are no specific guidelines about the order in which these qualifications should be achieved, I would strongly recommend completing them in the order laid out above. It is, however, possible for those with sufficient experience of performing inspection and testing to start with a Level 3 initial verification qualification, rather than starting at Level 2.

In truth, there is little to choose between the qualifications offered by the three awarding organisations, City & Guilds, EAL and LCL. City & Guilds qualifications are generally more popular due to the wider brand recognition and course availability, but you should consider all options before choosing a specific qualification.

SELECTING A TRAINING COURSE

Once you have selected a qualification, you then need to select a suitable training course. When selecting a training course it is essential to take your time and consider the wide range of options available. Usually most people will have an end goal in mind, i.e. the reason they are doing the course in the first place. Maybe you are applying for a job that requires you to hold a City & Guilds 2394; your natural instinct will be to book the cheapest course you can find or the one that is closest to your home. Sadly this is a mistake made by many people. You are a lot more likely to pass the course and achieve your goal by spending a little time on research and choosing the course that is right for you. Although the qualifications are the same, courses can differ massively; cheap courses can become very expensive if you fail to pass the exam.

When selecting your course, shop around. Below, I have included a checklist of typical things you should consider.

* Where is the centre situated?
* Is there good local accommodation and good transport links?
* How long is the course?
* Is it an evening course, block course or day release?
* What notes and handouts do they provide? (Get samples.)
* What equipment, books or materials do you have to provide?
* What is the expected trainer to delegate ratio?
* What is the background/experience of the trainer?
* How much practical content is there?
* What practical testing facilities do they have?
* Will you each have your own test equipment provided?
* How long has the centre been delivering this qualification?
* What is their current pass rate?
* What is their course cancellation policy?
* What is their re-sit policy?
* Do they have any reviews or testimonials you can see?
* Are discounts available if you book more than one course?
* Is funding or finance available for this course?
* What help/support is available after the course?

The better the answers you get to the above, the more you should be expecting to pay, but the better the learning experience you will have during the course and the better the chance you have of passing in the end.

All formal training courses will involve a detailed enrolment and course induction process. Make sure you provide any information to the centre as early as possible. This information will help the centre to best plan how they can meet your training needs – this is especially important for those with learning difficulties or health problems.

Part 2

In this part of the book I will explain in detail the subject matter that makes up the formal inspection and testing qualifications. This will give you an excellent guide to accompany your studies, whichever course you decide to undertake.

Most of the subjects referred to will relate directly to content in the *IET Wiring Regulations* (BS 7671), *IET Guidance Note 3* or the *IET On-Site Guide*. Where possible, I will direct you to the pages and chapters within these IET documents, allowing you to see the references first hand so that you can use these during your exam revision.

Part 2

CHAPTER 5

Inspection and testing safety

The most important part of any electrical work is safety and in this regard inspection and testing is no different. We must consider the safety of the inspector, other personnel working with the inspector or other trades, our customers and the general public and, finally, the impact that our work may have on the safety of the wider electrical installation and the building in which it is installed. During the inspection and testing process, the inspector is deemed to be in control of the electrical installation, therefore under the *Electricity at Work Regulations 1989* there are duties imposed on the inspector who, in this regard, is referred to as a 'Duty Holder'.

Later in this book I will discuss the subject of test equipment safety and the contents of *Health and Safety Executive Guidance Note* GS38. This useful and free-to-download document also contains good information with regards to general health and safety advice for those performing inspection and testing. Below I have expanded on the useful list from page 2 of GS38 'Things to consider when developing safe working practices'.

- **Control of risks while working** – It is important that prior to any inspection and testing work all hazards are identified, their risk calculated and, where necessary, suitable control measures put in place. Documented risk assessments record these activities and ensure that hazards, risks and control measures are clearly understood by all.
- **Control of test areas** – During the inspection and testing process it is essential, for safety reasons, that the inspector remain in complete control of the test area. Only those persons authorised to do so by the inspector should be able to enter; this is especially true where the test area may contain exposed live parts or other hazards. Controlling the test area may be achieved in simple cases by the use of suitable barriers, tapes and signs that restrict access. However in more complex environments where the simple measures are unlikely to prove effective, inspectors should consider additional precautions. These precautions may include: additional competent staff to supervise the test area, secure site fencing or the scheduling of work outside of normal hours. Where possible, the inspector should plan work activities to minimise the size of the hazardous work area and the amount of time taken to perform dangerous activities.
- **Use of suitable tools and clothing** – The inspector should, where possible, select tools and clothing appropriate to the possible hazards that may be encountered and in line with the control measures specified in the risk assessment detailed above. Examples include: suitably insulated handtools, flameproof clothing and protective footwear. It is also growing practice within the industry for those performing inspection and testing to use insulated rubber gloves and face shields, especially when undertaking live testing.
- **Use of suitable insulated barriers** – To improve safety while performing inspection and testing, the inspector may decide to fit insulated barriers to cover some exposed live parts while still giving access to perform the necessary inspections or tests. For example after removing a bus bar cover, the inspector may fit a temporary cover, made from insulating material and often transparent. This temporary cover may have small holes drilled to allow access for test probes but restrict the possibility of direct contact during the testing process.
- **Adequate information** – Prior to performing any inspection and testing activity, inspectors must ensure that they have enough information to

carry out their work in a safe manner. Sufficient information is required to perform a thorough risk assessment and the inspector must be fully informed about the nature and extent of the electrical installation, its use and history. If at any point the inspector feels that there is a lack of sufficient information to proceed safely, work should cease until such time that the information becomes available or can be established by the inspector through careful investigation. For example, where circuit details do not exist or cannot be provided by the customer, the inspector may need to postpone works until such time as all circuits can be traced. This work may need to take place outside of normal working hours to avoid danger to installation users or the public by the inadvertent disconnection of a misidentified circuit.

- **Adequate accompaniment** – Despite comprehensive safety precautions, electrical work still carries an inherent risk. It is therefore very important that in case an accident does occur the inspector has the ability to summon aid. Ideally every inspector would be accompanied at all times by another competent person, who is fully conversant with the risks posed by electricity, and is able to act quickly in the event of an emergency to make safe an electrical hazard and summon help without delay. The reality for many electricians is that lone working is the norm. It is therefore essential that this lack of accompaniment is considered within the risk assessment process and other control measures included to reduce the risk where possible. Other control measures may include: site sign-in sheets notifying others of your presence on site, calling in to the office before and after dangerous tasks to confirm safe completion, notifying first-aid staff on site of the activities being undertaken and the possible electrical dangers that may be encountered or the use of suitably trained customers' staff to supervise the inspector during safety-critical tasks.

- **Adequate space, access and lighting** – This topic often makes many electricians smile when they think of past occasions when they have performed inspection and testing, lying on their stomachs in a dark cramped loft space, with only a fading head torch to see by. We all have similar stories, but I'm sure all would agree that this is not really acceptable practice anymore. When planning any electrical activity we must consider how we are going to address these points. Having adequate space in which to work sounds like an obvious requirement, but how often do electricians work in cramped meter cupboards or lofts surrounded by cardboard boxes, old children's toys and the odd Christmas tree. It may be necessary prior to carrying out work to discuss the proposed work area with the customer and make arrangements where necessary for items to be temporarily removed. The requirement for proper access commonly relates to working at height but could equally apply to working in close proximity to fragile roof structures or ceilings, work in busy factory or production environments, access to hazardous areas or access to areas containing livestock. Again, access should be considered as part of the general risk assessment performed prior to commencement of the work activity, with suitable control measures specified and made available to the inspector. The requirement for adequate lighting is also essential to the safety of the inspector who must be able to see clearly at all times to perform safety-critical tasks. The provision of suitable temporary work lighting must also be considered at the planning stage.

While the above requirements for space, access and lighting apply to those performing inspection and testing activities, we must also consider the needs of other personnel, the customer and the general public. While we are performing inspection and testing we must plan for the continuing use of the installation by others. We must ensure that: suitable space is available for other users; access ways are not blocked; escape routes remain unobstructed; and fire exits remain clear, including access to first-aid boxes, fire extinguishers and emergency call points. Arrangements should be made to avoid or minimise the impact of inspection and testing work on building services, such as water, heating, lighting and power. Precautions should be taken to ensure that safety-critical systems remain effective, for example emergency lighting and fire alarm systems.

- **Precautions to prevent people not carrying out the testing coming into contact with exposed live parts** – This is achieved by applying a combination of the principles listed above. Contact with exposed live parts by persons not carrying out inspection and testing should be included as a hazard in your risk assessment. Risks should be assessed accordingly and control measures assigned. Typical control measures will include control of access to the work area, the use of suitable insulated barriers and proper planning of work activities to ensure live parts are exposed for the minimum amount of time.

The list above, while not exhaustive, will act as a good starting point for planning your inspection and testing work. Detailed planning accompanied by a comprehensive risk assessment are the essential foundations for safe inspection and testing.

CHAPTER 6

Safe isolation

I often refer to the subject of safe isolation as self-defence for electricians. Most electricians have had near misses, many have been injured and, sadly, some have been killed by a failure to safely isolate the electrical installation. As electricians, on a daily basis we are required to work on electrical installations that have been made dead, i.e. have been disconnected from their electrical supply. In many cases to perform our work we are required to touch parts of an installation that would normally be live, placing complete reliance for our safety on the integrity of the safe isolation process. Therefore rigidly adhering to the safe isolation procedure is the number one thing that electricians can do to improve the safety of themselves and others.

Because of its importance, the subject of safe isolation is addressed in the *Electricity at Work Regulations 1989*, which contain statutory requirements in relation to this area of electrical safety. Over the years guidance documents about the correct procedure for safe isolation have been many and varied, with each trade organisation offering its own take on how to implement the requirements of the *Electricity at Work Regulations 1989*. Recently, however, the Electrical Safety Council, under their new branding as Electrical Safety First, has produced a series of best practice guides that are available as free downloads from the Electrical Safety First website (www.electricalsafetyfirst.org.uk/electrical-professionals/best-practice-guides/). *Best Practice Guide 2* addresses the subject of safe isolation for low-voltage installations. The contents of this document have been agreed by a majority of the leading industry bodies and now form the basis for industry best practice in regards to safe isolation. I strongly recommend that all electricians read this document carefully and apply its guidance to their work. *Best Practice Guide 2* is also essential revision material to help those preparing for practical assessments and exams.

We will start by looking at the parts of electricity work regulations that address the subject of safe isolation. *Health and Safety Executive Guidance Document HSR 25*, available as a free download, elaborates on the requirements of the *Electricity at Work Regulations 1989* and is a good starting point for those new to the subject.

Regulation 14 of the *Electricity at Work Regulations 1989* states that:

> *No person shall be engaged in any work activity on or so near any live conductor (other than one suitably covered with insulating material so as to prevent danger) that danger may arise unless –*
> *a) It is unreasonable in all the circumstances for it to be dead; and*
> *b) it is reasonable in all the circumstances for him to be at work on or near it while it is live; and*
> *c) suitable precautions (including where necessary the provision of suitable protective equipment) are taken to prevent injury.*

Taken in isolation, the first paragraph of this regulation prohibits any person from working on or near live conductors. The hierarchy of controls, often used as a guiding principle for risk assessment, suggests that the best control measure is to remove the hazard in the first place. Therefore, common sense would suggest that wherever possible normal working practice for electricians should be to avoid working on or near uninsulated conductors. Unfortunately there are occasions when this type of work is unavoidable; inspection and testing is generally considered to be one of these areas. When working on or near uninsulated conductors we must consider points a) to c) above.

Point a) applies the stipulation that it must be 'unreasonable' in all circumstances for the conductor to be dead while work is carried out. The duty holder must be able to make a case that isolating the conductor is 'unreasonable'. There are very few types of electrical work where this would be true. Following on from point a), point b) adds a further stipulation that work on or near the conductor while it is live is indeed 'reasonable'. This second point adds common sense to the regulation – just because it is not reasonable to make something dead does not mean that it is reasonable for you to work on it while it is live. These two tests must be applied in unison before live working can be considered. The third and final point, point c), suggests that once we have met the requirements of points a) and b) and decided to work live, we must take suitable safety precautions to prevent injury during the works. Whenever you are required to carry out any work on or near live uninsulated conductors, you must consider the legal implications of regulation 14; this is especially true when planning inspection and testing work.

Regulation 13 of the *Electricity at Work Regulations 1989* states that:

> *Adequate precautions shall be taken to prevent electrical equipment, which has been made dead in order to prevent danger while work is carried out on or near that equipment, from becoming electrically charged during that work if danger may thereby arise.*

Regulation 13 relates more closely to the topic of safe isolation. It stipulates that precautions are taken to prevent equipment that has been made dead from becoming live while work is carried out. Compliance with this regulation is usually ensured by the incorporation of a suitable 'lock off' method as part of the overall safe isolation procedure.

Regulation 12 paragraph 2 gives a definition of the term 'isolation', which it explains is the 'disconnection and separation of the electrical equipment from every source of electrical energy in such a way that this disconnection and separation is secure'. This definition further supports the need to 'lock off' circuits in a secure manner.

To help us comply with legal obligations, *Best Practice Guide 2* details safe isolation procedures that we can follow. Below, I will attempt to summarise and, where necessary, explain this safe isolation procedure. Being able to describe and carry out the safe isolation of electrical installations is always a key part of any course assessment. It is for this reason that practice and revision of this topic is highly recommended.

PREPARATION FOR SAFE ISOLATION

Before isolating any circuits it is essential that careful consideration is given to the safe isolation process. You must consider the implications of performing the safe isolation for yourself and others, how the isolation is to be performed and the suitability of equipment or devices used for safe isolation. It will also be necessary to notify any parties affected by the isolation and, where necessary, seek their written permission. Proper planning will ensure the process proceeds in a smooth and, most importantly, 'safe' manner.

Turning off the electricity supply to safely isolate parts of the electrical installation will inevitably cause a certain amount of disruption. Prior to disconnection of the supply, we must take measures to minimise this disruption. We must consider the implications to ourselves, other personnel, the customer, their employees and the general public. We must also consider the impact of this isolation on other essential building services. In some cases temporary supplies may be required for essential equipment, or the work may need to be carried out at a time chosen to cause the least disruption.

To perform a safe isolation we first need to identify an isolation point. This point of isolation usually takes the form of a protective device, isolator or switch situated 'upstream' on the supply side of the conductors and/or equipment that require isolation. In any installation the preferred point of isolation is always the main switch. Disconnecting the complete installation from the electricity supply minimises the chances of danger resulting from a misidentified circuit or cross connections between circuits, such as a 'borrowed' neutral. Where isolation of the complete installation is not practical, it is acceptable to isolate part of the installation, but great care must be taken to avoid danger.

When selecting the point of isolation a certain amount of investigation may need to be carried out to ensure that the device selected does in fact control the supply to the circuits or equipment being isolated. This is especially true for older installations where labelling of devices is likely to be poor or circuit diagrams and charts are out of date.

Once we have selected the point of isolation we must ensure that this device is suitable for the task. Is the device capable of being locked in the 'off' position and do we have the necessary 'lock off' adapters and locks for this device? Is this device constructed to the correct British or harmonised standard, which signifies the device is suitable for the purpose of isolation? The *IET Wiring Regulations* (BS 7671) contains table 53.4 on page 153, which is also included in *Best Practice Guide 2* and lists current isolation and switching device standards. The table details which devices are suitable for isolation and which are not. The *Best Practice Guide 2* also reminds us that some older devices, for example BS 3871 circuit breakers, are unlikely to be suitable for isolation. Also remember that in TT or IT systems, single pole isolation is not acceptable; the chosen isolation device must be able to disconnect the neutral too.

To perform safe isolation we require a certain amount of tools and equipment, in addition to the personal protective equipment and safety equipment specified in your risk assessment. You will require a 'lock off' adapter that fits your chosen isolation device, a safe isolation sign/notice, a padlock with only one key or a combination lock with a unique code, a suitable voltage indicator or test lamp (details below) and a known voltage source or proving unit. For isolations involving more than one person, it is also recommended that a multi-lock hasp is used to allow each person working on the isolated installation to fit their own individual lock. Therefore the multi-lock hasp can only be removed following the removal of all individual locks. Prior to attempting safe isolation it must be confirmed that all equipment mentioned above is available, in good working order and suitable for the planned work.

Arguably, the most important piece of equipment required to perform the safe isolation process is an approved voltage detector or test lamp. While voltage can be detected with multi-meters, non-contact voltage detectors (voltsticks) and other similar equipment, the use of these devices has in the past caused accidents, for example when a multi-meter has been set to the incorrect range or leads have been connected to the wrong terminals. Therefore, a two-pole voltage detector complying with BS EN 61243-3 is the safest way to confirm that a circuit is safely isolated.

Health and Safety Executive Guidance Note GS38 contains detailed best practice recommendations for the safety of test equipment used in low voltage installations. Below, I have picked out some of the key *GS38* guidance (from GS38 pages 4 to 6) in relation to test probes, clips, leads and two-pole voltage detectors.

- They should comply with BS EN 61010 or BS EN 61243-3.

- They should be marked with the correct overvoltage category rating (CAT II, III or IV) and maximum voltage.

- Probes should have **finger barriers** or be shaped to guard against accidental hand contact with the live parts.

- Test probes and clips are insulated to leave an exposed metal tip not exceeding **4 mm**, however it is strongly recommended that this is reduced to **2 mm** or less, or that spring-loaded probes are used.

- When used with a multi-meter, test probes should be fused.

- Test leads should be adequately **insulated** and **sheathed** to protect against mechanical damage.

- Test leads should be coloured so that one lead can be **easily distinguished** from the other.

- Test leads should be **flexible** and of **sufficient capacity**.

- Test leads should be **long enough** for the purpose, but **not too long** that they are difficult to use.

- Test leads should **not have accessible exposed conductors** other than the probe tips, or have live conductors accessible to a person's finger if a lead becomes detached from a probe or equipment when in use.

FIGURE 6.1   Examples of test leads that do not comply with GS38

In the list above, I have mentioned the overvoltage category rating of the test equipment. This is an important consideration that requires further explanation. If we are measuring a voltage, of say 230 VAC, you may imagine that a test meter and leads rated to 230 VAC would be sufficient; this is not necessarily the case. Before taking a measurement on circuits that have the potential to be live, even if we believe they are dead, we must use equipment that is rated, not for the expected voltage, but for the maximum voltage that could be present.

All electrical installation can be subjected to overvoltages, surges or spikes in the voltage that may last only a few milliseconds or several seconds. These overvoltages could be due to a variety of reasons, both internal and external to the installation. Typical examples include: faults in the distribution network, local lightning strikes, high voltage switching or the switching of large inductive loads. These events can cause overvoltages of several thousand volts, therefore if we were unlucky enough to be taking a measurement at the same time as the overvoltage occurs, our safety would depend on the test equipment and leads being able to safely cope with the situation.

To help us select the correct test equipment for the type of installation we are working on, we use Category (CAT) ratings combined with the maximum nominal voltage to the installation. Essentially, the closer we are working to the origin of the installation the higher the overvoltages could be and the higher CAT rating we require for our equipment. Measurements taken at the origin on the installation up-stream of the main distribution board are deemed to require protection to CAT IV. Measurements usually taken at the origin include: confirmation of supply polarity, external earth fault loop impedance and prospective fault current. CAT III generally refers to the fixed electrical installation and includes: distribution boards, fixed wiring and fixed electrical equipment. CAT II covers equipment connected to the fixed installation, such as appliances and portable tools. The *IET Wiring Regulations* (BS 7671) have further information on page 97.

Failure to ensure the use of correctly rated test equipment could lead to serious injury, or even death, in the event of an overvoltage. Great care should be taken to understand the category of installation you are going to be working on and the suitability of test equipment for that environment before any measurements are attempted.

Before performing any isolation activity, it is essential that we consult the customer, and any other interested parties, to clearly agree the planned isolation. All parties must be aware of when the isolation will take place, which parts of the installation will be isolated, what impact this will have on those present and what additional safety measures have been put in place. Often the agreed safe isolation process will be documented in a risk assessment, method statement or permit to work.

THE SAFE ISOLATION PROCESS

Following the successful completion of the preparations detailed above, and subject to agreement by the customer and other interested parties, we are now able to start the safe isolation process.

The safe isolation process begins by switching off at the point of isolation identified earlier, and making the isolation secure using a suitable 'lock off' adapter and lock. The unique key or key code for this lock must be retained only by the person performing the isolation, ensuring that the isolated circuit remains under their control at all times. A safe isolation notice must now be fitted at the point of isolation to make others aware of the isolation details. The safe isolation notice will usually include a

FIGURE 6.2   A safe isolation notice

warning symbol, details of the circuit that has been isolated, the name of the person who has performed isolation with contact details and the reason for this isolation.

Once the safe isolation notice has been fitted, the person performing the safe isolation should move to the point of work, where the installation or circuit to be worked on must be proved 'dead' prior to commencement of works. Until conclusively proved 'dead' we must assume that the circuits and equipment to be worked on are still 'live' and therefore all 'live working' precautions, including suitable personal protective equipment, must be adhered to.

- Prior to removing any covers, the approved voltage detector should be thoroughly checked for suitability and to identify any damage or defects. It must then be proven on a known voltage source, usually in the form of a dedicated proving unit, to confirm correct operation.
- Adhering to 'live working' precautions, the person performing the safe isolation should then remove any covers required to access the point of work. The approved voltage detector is then used to test for voltage between all live parts and between all live parts and earthed parts (exposed or extraneous conductive parts). It is also necessary to test for hazardous voltages at any other terminals where these voltages may be present such as control circuits, DC circuits or switch wires. Great care must also be taken where multiple sources of electrical energy exist, e.g. batteries, generators, capacitors or PV modules.

As a minimum, the below measurement of voltage will be made:

No voltage should be detected at the point of work at any time during this process. If any voltage is detected then the isolation should be halted and all covers immediately replaced, until such time as the failure in the isolation process can be identified and corrective action taken.

- Finally, to confirm that the voltage detector has not failed during the isolation process, we must reprove the voltage detector on the known voltage source. Only when the voltage detector is confirmed to be working can we accept the results of the above tests and consider the point of work safely isolated. It is often advisable to re-confirm the safe isolation periodically, for example after a lunch break or when returning to site after a trip to the wholesalers. This may seem like paranoia, but better safe than sorry.

TABLE 6.1 Single phase (three tests)

Test between	Expected voltage
Line and Neutral	0V
Line and Earth	0V
Neutral and Earth	0V

TABLE 6.2 Three phase (ten tests)

Test between	Expected voltage
L1 and Neutral	0V
L2 and Neutral	0V
L3 and Neutral	0V
L1 and Earth	0V
L2 and Earth	0V
L3 and Earth	0V
L1 and L2	0V
L1 and L3	0V
L2 and L3	0V
Neutral and Earth	0V

The above three-step process forms the basis for safe isolation: prove the voltage detector, prove 'dead' and then reprove the voltage detector. This process is at the heart of all safe isolation activities and a thorough understanding of this process is key to your exam preparation. Practising this process until it becomes second nature will also ensure a confident performance in your practical assessments and will help to demonstrate your competence to the assessor.

CHAPTER 7

Information required for inspection and testing

In addition to the health and safety requirements covered in the previous chapter, before carrying out any inspection and testing activity it is essential that the inspector has all the necessary information required. This information comes in many forms but basically breaks down into three types: statutory, non-statutory and information about the installation.

The first and most important things to understand are the terms **statutory** and **non-statutory**. Statutory is the term used to describe something that is a legal requirement and therefore must be complied with by law. Something that is non-statutory, however, usually exists simply to provide guidance and is not a legal requirement. These definitions are a huge simplification, but will do fine for our purposes.

Typical examples of statutory documents associated with inspection and testing include:

- The *Electricity at Work Regulations 1989*
- The *Health and Safety at Work Act 1974*
- The *Electricity Safety, Quality and Continuity Regulations 2002*
- The *Building Regulations 2010.*

Typical examples of non-statutory documents associated with inspection and testing include:

- The *IET Wiring Regulations* (BS 7671:2008)
- *IET Guidance Note 3*
- *IET On-Site Guide*
- *Electrical Safety First – Best Practice Guide 2*
- *HSE Guidance Note GS38*
- *HSE Guidance on the Electricity at Work Regulations HSR 25.*

Both statutory and non-statutory documents are important to us and it is essential that we know the difference between them.

INITIAL VERIFICATION INFORMATION

As discussed in the Introduction to this book, initial verification is the type of inspection and testing applied to new installations and also covers additions and alterations to existing installations. The general requirements for initial verification are laid out in section 610 of the *IET Wiring Regulations* (BS 7671) on page 197. Remember that the purpose of initial verification is to confirm, by inspection and testing, that the new installation complies with the requirements of BS 7671. For additions and alterations we must also confirm that they do not impair the safety of the existing installation. As part of the initial verification it will be necessary to compare our results with relevant criteria such as data from BS 7671 or calculation data provided by the designer.

In order to carry out initial verification it will be necessary for you to have certain information made available to you. Regulation 610.2 requires information to be made available about three areas:

- The assessment of fundamental principles (section 131), which covers protection for safety.
- General characteristics (sections 311 to 313), covering maximum demand, diversity, conductor arrangements, earthing systems and supply characteristics.
- Diagrams and documentation (regulation 514.9.1), which includes the information contained in a distribution board schedule.

PERIODIC INSPECTION AND TESTING INFORMATION

Periodic inspection and testing is carried out on existing installations to establish whether the installation is a satisfactory condition for continued service. The general requirements for periodic inspection and testing are laid out in section 621 of the *IET Wiring Regulations* (BS 7671) on page 203 and break down into four main areas of safety provided by the installation.

- The safety of persons and livestock against the effects of electric shock and burns.
- Protection against damage to property by fire and heat arising from an installation defect.
- Confirmation that the installation is not damaged or deteriorated so as to impair safety.
- The identification of installation defects and departures from the requirements of the *IET Wiring Regulations* (BS 7671) that may give rise to danger.

The above four areas form the basis for all periodic inspection and testing and it is important that we keep them in mind at all times during the inspection testing process.

In order to perform periodic inspection and testing the inspector will need access to a wide range of information. In addition to the statutory and non-statutory documents mentioned earlier in this chapter, the inspector will also require details relating to the original initial verification of the installation, including the electrical installation certificate. Details of previous periodic inspection and testing carried out on the installation, along with any maintenance records, are also essential.

Whether performing initial verification or periodic inspection and testing, the more information the inspector has prior to carrying out the inspection and testing activity, the better prepared they will be. Having complete information is essential to performing risk assessments and managing the health and safety of yourself and others during the inspection testing activity. Detailed information about the installation is also invaluable during the inspection and testing process, allowing you to compare the results of inspection and testing with previous information, to either establish compliance with current regulations for new installations or identify deterioration and trends in existing installations. Remember, if sufficient information is not available to the inspector to allow the work to proceed in a safe manner, work should not be carried out until suitable information can be obtained or prepared.

CHAPTER 8

The difference between initial verification and periodic inspection and testing

When asked to describe the differences between initial verification, periodic inspection and testing, many people will simply state that initial verification is for new installations and periodic inspection and testing is for old ones. This is true, but it fails to capture the significant differences between the two processes.

Initial verification is a very neat and orderly, almost clinical, process, carried out on a new installation throughout its construction and on its completion. Because this is a new installation the inspector will have all information required and ready access to all parts of the installation. Initial verification works can be sequenced to allow inspection and testing to take place in a methodical, step-by-step manner. The inspector may even be the person who has constructed installation, so will have a detailed knowledge of the extent of the installation and its layout.

Now contrast the above with **periodic inspection and testing**. In this process the inspector must determine the condition of an existing installation, often while it is still being used by the customer. This installation may have been designed to comply with an old edition of the wiring regulations and since its original installation may have been subject to several alterations or additions. Often these alterations and additions may have been performed by DIY electricians, with little regard for wiring regulations. Much of the installation may also be inaccessible to the inspector, concealed within the fabric of the building, at height or within inaccessible roof spaces, for example. Therefore, it is clear that the realities on the ground make it impossible to perform a full initial verification type process on existing installations.

Another important factor to consider with existing installations is the fact that they are usually energised and fully working when the inspector first encounters them. This is the reverse of the initial verification process, where the installation will usually be 'dead' and only be put into service following successful completion of the initial verification. Great care and planning is required to establish a safe sequence of work for the periodic inspection and testing of existing installations that are currently in use.

It is also important for the inspector and the client to understand that periodic inspection and testing is not a maintenance activity, where the installation is repaired as we go along. The inspector must report only on the current condition of the installation. The client can then plan remedial works after the report's completion based on the inspector's recommendations. This separation between inspection and testing and remedial works is essential if we are to be able to gauge the true condition of the installation's safety. If installation defects are simply repaired and never recorded it is impossible to track patterns of failure, establish trends or plan future inspection and testing activities with any accuracy.

You can see from the above that periodic inspection and testing requires a significantly different approach from that employed for initial verification. In the rest of this chapter I will talk in detail about the periodic inspection process and help you to understand this often misunderstood part of inspection and testing.

REASONS FOR PERIODIC INSPECTION AND TESTING

The main statutory justification for periodic inspection and testing comes from the *Electricity at Work Regulations 1989* and, while it does not specifically require inspection and testing, it is often quoted because inspection and testing are necessary to establish the need for 'maintenance', which is a requirement – see regulation 4(2) below.

Regulation 4(2) of the *Electricity at Work Regulations 1989* states that:

As may be necessary to prevent danger, all systems shall be maintained so as to prevent, so far as is reasonably practicable, such danger.

There are a few interesting terms used here that help us understand the maintenance requirement. The term 'as may be necessary to prevent danger' places a legal duty on us to establish the correct frequency for maintenance activities and is quite open-ended as long as we ensure that danger is prevented. The term 'all systems' refers to all electrical systems; naturally this includes the electrical installations in buildings and also the electrical equipment within the buildings. The final term of interest is 'reasonably practicable'. This gives us the ability to apply a little common sense to the requirement; we must determine what is reasonable and practical. So we must make a decision that balances the requirement to prevent danger against what is reasonable and practical in a real-world environment.

In addition to the above, there are many reasons why a customer may wish to have periodic inspection and testing performed on their installation. I have summarised these reasons below:

- **Change of ownership** – when a property changes hands the seller or potential new owner may wish to establish the condition of the electrical installation. A satisfactory Electrical Installation Condition Report

(EICR) may also be required by the mortgage company before they are willing to lend money to purchase the property.

- **Insurance** – many insurance policies include a requirement for regular periodic inspection and testing; also that, where necessary, remedial works are carried out to maintain the installation in a safe condition.
- **Change of use** – before a building is converted or adapted it is essential that the condition of the existing installation is established. The designer of the new installation can then use the Electrical Installation Condition Report (EICR) to inform the design process and ensure that the new works do not impair the safety of the existing installation.
- **Increased electrical loading** – before adding new circuits or equipment to an existing electrical installation, it may be necessary to perform periodic inspection and testing to establish the impact of this increased electrical loading on the existing installation. For example, before the electrical supply for a temporary Portakabin is connected to the existing electrical installation in a school, it will be necessary to establish that this increase electrical loading will not impair the safety of the existing installation or the supply.
- **Licensing** – many premises are subject to licensing requirements. These may include, for example, pubs, entertainment venues, caravan sites, petrol stations and houses in multiple occupation (HMOs). Licensing authorities may require regular periodic inspection and testing and, where necessary, remedial works are carried out to maintain the electrical installation in a safe condition.
- **Where damage is suspected** – when the electrical installation has been subject to, for example, fire or flooding, periodic inspection and testing may be necessary to confirm the safety of the electrical installation or identify damage that must be rectified.

In addition to any of the reasons stated above, all electrical installations are subject to deterioration, and the *IET Wiring Regulations* (BS 7671) recommends, in regulation 135.1 on page 22, periodic inspection and testing for all electrical installations.

EXTENT AND LIMITATIONS

Once the client has decided that they require periodic inspection testing to be carried out on the electrical installation, the first conversation they will have with the inspector will usually be a meeting to agree 'extent and limitations'. The term 'extent' is used to describe which parts of the electrical installation will be subjected to inspection and testing. Many clients simply assume that the periodic inspection and testing process will cover everything, but in many cases this level of inspection and testing is not necessary or not even possible. For example, when conducting a periodic inspection and test of the electrical installation in a flat, the extent is usually limited to the electrical installation within the flat and does not extend to the lighting in communal areas or the wiring of the distribution circuit that connects the consumer unit in the flat to the electricity supplier's meter in the landlord's meter cupboard on the ground floor. These areas may, however, be covered within the extent of a periodic inspection and test commissioned by the landlord.

Unlike initial verification, it is common practice when performing periodic inspection and testing to use sampling rather than attempting to inspect and test the complete electrical installation. Sampling methodology is covered in detail in *IET Guidance Note 3* (page 76) and the aim of this process is to select a representative sample, which will allow us to accurately gauge the condition of the electrical installation without the need for a 100 per cent check. For example, we may decide to sample 50 per cent of lighting circuits. Details of the proposed sampling strategy are also recorded as part of the 'extent' agreed prior to starting periodic inspection and testing. When performing inspection and testing on the sample circuits, the inspector may discover defects or non-compliances. In these cases it is usually expected that the inspector will increase the sample size to establish the extent to which the defect or non-compliance occurs throughout the rest of the installation.

Once we have established the 'extent' of the periodic inspection and testing activity, we must then look at 'limitations'. Limitations is the name given to the limits imposed on the inspection and testing activity by practical considerations. For example, in a factory it may not be possible to isolate some production equipment, limiting the range of testing that can be carried out on certain circuits, or certain rooms in a bank may be off limits to the inspector for security reasons. These are limitations known to the inspector beforehand and are recorded as part of the 'extent and limitations' declaration on the Electrical Installation Condition Report (EICR).

The 'extent and limitations' can be used to tailor the inspection and testing process to cover anything from the complete installation to maybe just one circuit or even one piece of equipment. It is vitally important that the 'extent and limitations' are fully understood by all parties and the implications too are made clear. In addition to agreement between the inspector and the client, it may be necessary to seek agreement from other interested parties, such as licensing authorities, insurance companies, mortgage lenders or landlords. Once agreed, the 'extent and limitations' are formally recorded in section D of the Electrical Installation Condition Report (EICR). This allows all future readers of the report to fully understand to which parts of the electrical installation the findings of the report refer.

Once the inspection and testing process begins, the inspector may find that there are limitations imposed that were not envisaged during the planning process and therefore not included in the original 'extent and limitations'. Section D of the Electrical Installation Condition Report (EICR) contains a section for these 'operational limitations' to be recorded along with suitable reasons for these limitations.

OBSERVATIONS

When performing periodic inspection and testing you will naturally discover deficiencies in the installation that you feel should be recorded in the Electrical Installation Condition Report (EICR). These 'observations' are recorded in section K of EICR.

Observations generally cover two categories of deficiency: **defects** and **non-compliances**. A defect, as the name suggests, refers to a part of the installation that is 'defective', i.e. has become damaged or worn to a point where it affects the continuing safety of the electrical installation. A non-compliance, however, refers to a part of the installation that fails to comply with the latest edition of the *IET Wiring Regulations* (BS 7671). Non-compliances are likely to be very common in older installations, because they were installed to earlier editions of the wiring regulations. Where installations were installed to earlier versions of the wiring regulations, it is only necessary for the inspector to record observations where the non-compliance could give rise to danger.

When recording observations it is important that the language used is concise and expressed in a way that can be clearly understood by the client. The inspector should avoid overly technical terms like Zs or PFC, with the emphasis on accurately describing the deficiency and not what remedial work is required. Many inspectors will also include photographs to aid the client's understanding. Observations should be laid out logically and usually numbered sequentially.

To help the client prioritise remedial works and to understand the significance of each observation, classification codes are used. Each observation is assigned a classification code by the inspector. There are three codes: **C1**, **C2** and **C3**.

- **Code C1** – Danger present, risk of injury, immediate remedial action required.
- **Code C2** – Potentially dangerous, urgent remedial action required.
- **Code C3** – Improvement recommended.

On the very rare occasion when the inspector is not able to classify an observation without further investigation that is beyond the scope of the extent and limitations previously agreed with the client, the code FI may be used to communicate to the client that 'Further investigation is required without delay'.

If the inspector encounters any situation that poses an immediate danger, the inspector should inform the client without delay and in writing. Where possible, action should also be taken immediately to remove the danger until such time as remedial works can be carried out. As mentioned earlier in this book, during the inspection and testing activity the inspector is regarded as a Duty Holder under the *Electricity at Work Regulations 1989* and as such they have a legal responsibility to ensure the safety of themselves and others. Therefore the inspector does have a legal duty to make the installation safe.

It is the job of the inspector to assign the correct code to each observation. This is a difficult task and many new inspectors are often tempted to use the C1 code for almost everything. In fact C1 is not applicable in most cases and is only used for the most serious deficiencies, for example where live parts are exposed and accessible to the user. Most observations will generally only be a C2 or C3.

Assigning suitable classification codes to our observations has long been a problem within the electrical industry and many experienced electricians still disagree about which observation requires which code.

To bring a little clarity to the subject, Electrical Safety First have produced a free best practice guide on the classification codes, containing many examples of C1, C2 and C3 deficiencies. *Best Practice Guide 4 (issue 4)* is essential reading for all those new to the subject of periodic inspection and testing and is available from their website (www. electricalsafetyfirst.org.uk/electrical-professionals/best-practice-guides/). The contents of the guide have been agreed by most major industry bodies and, therefore, we can use this guide to calibrate our own practice, ensuring standardisation with the wider industry. Those revising for periodic inspection and testing exams or practical assessments should spend time on assigning classification codes; this is always an area tested by the examiners.

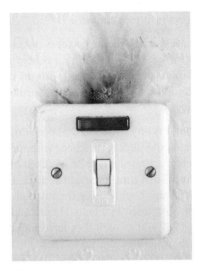

FIGURE 8.1   How would you code the switch above?

Once all observations have been recorded and suitable classification codes assigned, the inspector must then summarise their findings in section E of the EICR. This summary should detail the general condition of the electrical installation with regards to electrical safety, i.e. how well the installation meets the requirements of Part 4 (Protection for Safety) of the *IET Wiring Regulations* (BS 7671). Section E also requires the inspector to make an overall assessment of the installation in terms of its suitability for continued use. The inspector must classify the installation as either satisfactory or unsatisfactory.

If any observations have been classified C1 or C2, or FI then the installation's condition must be recorded as unsatisfactory. Where no

C1 or C2 observations are made, the inspector will usually deem the installation condition as 'satisfactory'. Section F requires the inspector to recommend a date for the next inspection, assuming all remedial actions have been carried out. *IET Guidance Note 3* contains a table (Table 3.2), on page 73, that gives guidance for the recommended initial frequencies for periodic inspection and testing (i.e. the suggested maximum length of time to first periodic inspection and testing, specified by the designer). For subsequent periodic inspection and testing, the maximum time between inspections is specified by the inspector. There is no set re-test period, so this decision is based on the condition of the installation and other factors, such as external influences. The recommended re-test date is recorded in section E of the EICR and also on a periodic inspection and testing notice fixed near the origin of the installation.

CHAPTER 9

Inspecting an electrical installation

Inspection is the term used to describe the activity of using our senses to establish the condition of the electrical installation.

Reference is often made to visual inspection, however **sight** is only one of the senses we come to rely on during the inspection process. The inspector may use **touch** to feel for sharp edges inside a newly threaded piece of conduit or to check the tightness of connections. **Smell** is often the first indication of an overheating motor or burned out terminal. Your **hearing** can be used to detect a worn bearing or the clunk of a circuit breaker tripping out.

While sight is undoubtedly the most common sense used during the inspection process and can be used to detect anything from a missing label to exposed live parts, an inspector who relies solely on sight limits their ability to detect a full range of faults. Talking about the senses used during inspection may seem obvious, but it is an area often used by examiners in formal exams. You should prepare some examples as part of your exam revision. Remember the inspector will be looking for different things when performing initial verification compared with periodic inspection and testing, so you will need examples specific to each subject.

INITIAL VERIFICATION INSPECTION

When inspecting new installations, inspection should always precede testing and the electrical installation should be safely isolated from its supply. The *IET Wiring Regulations* (BS 7671) detail the inspection requirements in section 611 on page 197.

There are three areas listed in BS 7671 that require inspection as part of initial verification:

- All equipment complies with appropriate British or harmonised standards, in accordance with section 511 of BS 7671.
- All equipment is correctly selected and erected in accordance with the requirements of BS 7671.
- No equipment is damaged or so defective as to impair safety.

Regulation 611.3 on page 197 of the *IET Wiring Regulations* lists the minimum items to be inspected. However this is not a specific checklist with items that can be easily confirmed and checked off; the list contains general principles, such as connection of conductors or methods of protection against electric shock.

To aid our understanding of the items contained in regulation 611.3, *IET Guidance Note 3* (section 2.5.2 on page 18) expands on each list item. Later, in section 2.5.3 on page 27, *Guidance Note 3* also contains a detailed list of example checklist items, allowing us to get a comprehensive understanding of the depth expected from an initial inspection. Amendment 3 of the *IET Wiring Regulations* (BS 7671) has introduced a new 'schedule of inspections' form (page 422), which is used to record the initial inspection. This form is significantly easier to complete than its predecessor, which was based solely on the items listed in regulation 611.3. The new form is designed to include items that can be easily understood and checked. Many items are also accompanied by a regulation number so that the inspector can cross-reference where necessary.

When preparing for your initial verification exam or practical assessment, you should study the inspection items in detail and also practise completing the 'schedule of inspections' for different installations. During exams you may be asked to list, for example, five items that should be inspected for a new metal conduit system before installation of cables. Many candidates lose marks by using incorrect terms or simply by choosing bad examples of items to inspect. Instead we can use the conduit systems checklist on page 30 of *IET Guidance Note 3*, which contains a list of suitable items, to help us answer this type of question. I always recommend that candidates preparing for exams spend time on this type of question during their revision.

PERIODIC INSPECTION

When performing periodic inspection of existing installations, the purpose of the inspection is to confirm whether the installation is safe for continued service. This inspection should be carried out with the minimum possible dismantling of the installation and will usually involve a sampling strategy to aid in this regard. Unlike initial inspection, which always takes place before testing, it is likely that periodic inspection will be carried out at different stages throughout the process, as parts of the installation become available. For example it may be necessary to fully inspect and test one circuit, before the inspector can then move to the next circuit and so on. This more fragmented approach is normal with periodic inspection, especially when the installation is in use during the inspection. The inspector must plan works carefully and ensure safety at all times.

To aid the inspector, the acronym SADCOWS is often used to remind us of the things we are evaluating during a periodic inspection.

- Safety
- Ageing
- Damage
- Corrosion
- Overloading
- Wear and tear
- Suitability.

The above is not an exhaustive list, but it does help to illustrate the differences between periodic and initial inspections. During a periodic inspection, the inspector must consider many more factors and each

FIGURE 9.1   Remember the acronym SADCOWS

must be assessed to decide at what point they render the installation unsafe for continued service. When does the age of an installation make it unsafe? How much wear is acceptable? These are not easy questions to answer and this is why periodic inspection is such a difficult art.

Amendment 1 of the *IET Wiring Regulations* introduced a new 'electrical installation condition report' form and a new 'condition report inspection schedule' (page 432) used to record the outcome of a periodic inspection. The new inspection schedule lists the numerous parts of the electrical installation that require inspection and is very similar to the form now used for recording initial inspections mentioned above. When completing the condition report inspection schedule, the inspector must choose from eight outcome options.

- ✓ to indicate that the condition of this item is acceptable.
- **C1** to indicate that the condition of this item is unacceptable (Danger present, Risk of injury, Immediate remedial action required).
- **C2** to indicate that the condition of this item is unacceptable (Potentially dangerous, Urgent remedial action required).

- **C3** to indicate that improvement is recommended for this item (Improvement recommended).
- **FI** to indicate that further investigation is required without delay.
- **N/V** (Not Verified) to indicate that the inspector was not able to verify the condition of this item.
- **LIM** (Limitation) to indicate that this item falls outside of the scope of this report due to a limitation detailed in section D of the EICR.
- **N/A** (Not applicable) to indicate that list items are not applicable to this installation.

As you can see from the above, completing the condition report schedule is a lot more complicated than completing the initial verification schedule, which has only two outcomes, a tick or N/A. Remember that these outcomes should also tie in with the information on the EICR. So, for example, if we have put a C1 in the outcome box for 'Identification of conductors', then we would expect to see an observation explaining this in section K.

As for initial verification above, those preparing for periodic inspection and testing exams or practical assessments should include study of the items contained in the condition report inspection schedule, including practical examples of each. You should also study the eight outcomes detailed above and be able to use them correctly when completing the inspection schedule.

IP CODES

The *IET Wiring Regulation* (BS 7671) abbreviations list, on page 41, states that IP stands for International Protection. However, in other publications, IP is often defined as Ingress Protection or even Index of Protection.

Whichever definition you prefer, IP codes relate to the protection offered by enclosures, as defined in BS EN 60529. The IP code tells us how well a person is protected against access to hazardous parts inside the enclosure and how well the enclosure is protected against the ingress of solid foreign objects and the ingress of water. IP codes allow us to quantify the characteristic of enclosures, which can then be specified in relation to the external influences present. Those writing standards such as the *IET Wiring Regulations* (BS 7671) can use IP codes to insist on a certain degree of protection for enclosures in specific environments, for example IPX4 for zone 1 in a bathroom.

Designers can use IP codes to help them select the right equipment to cope with specific external influences, for example specifying an IP55 luminaire in a carpenter's workshop where there is a moderate amount of wood dust in the air (AE5). As inspectors it is our job to confirm that both the IP requirements of the wiring regulation and those specified by the designer have been met and that enclosures offer the necessary levels of protection required. To be able to inspect enclosures, therefore, it is necessary for inspectors to have a good understanding of IP codes and know which enclosure should be used where.

IP codes usually take the form of two letters, IP, and two numbers, for example IP44. The first number relates to the protection provided by the enclosure against persons accessing live parts and also the protection provided against the ingress of solid foreign bodies. It ranges from 0 to 6. The second number relates to the enclosure's ability to protect against the ingress of water. This number ranges from 0 to 8. The higher the numbers are, the greater protection is offered, therefore an IP00 enclosure would offer no protection and an IP68 enclosure would offer the highest protection.

Unfortunately IP codes are not clearly defined in the *IET Wiring Regulations* (BS 7671) or *IET Guidance Note 3*. They are, however, addressed in appendix B of *IET Guidance Note 1* (page 137).

Table 9.1 summarises the meaning of the main IP codes used. However, for a detailed definition please refer to BS EN 60529.

As mentioned in Table 9.1, some IP codes may contain the letter X. This signifies that the code does not specify specific protection in this area. For example, a regulation may specify an IP2X code for an enclosure that requires protection access by fingers. As this regulation does not concern protection against water ingress no protection against water ingress is specified and an X is used for the second number in the IP code. X codes are generally only used in regulations, and manufacturers tend to use the normal IP code format.

Some older enclosures may have been designed to a slightly different standard. To allow for this the IP code system uses two Xs and an additional letter at the end of the code. The *IET Wiring Regulations* (BS

Table 9.1 Summary of the meanings of the main IP codes

Code	First number (Protection against solids)	Second number (Protection against water)
0	No protection	No protection
1	Objects ≥ 50mm (Back of hand)	Vertically falling drops
2	Objects ≥ 12.5mm (Finger)	Vertically falling drops, enclosure tilted at 15°
3	Objects ≥ 2.5mm (Tool)	Water spraying at up to 60°
4	Objects ≥ 1mm (Test pin)	Splashing water
5	Dust protected	Water jets
6	Dust tight	Powerful water jets
7	N/A	Temporary immersion in water
8	N/A	Continuous immersion in water
X	Not specified	Not specified

Table 9.2 Additional codes

IPXXA	Protected against access with back of hand. A sphere ≥ 50mm diameter has adequate clearance from live parts.
IPXXB	Protected against access with a finger. A test finger ≥ 12mm diameter and 80mm long has adequate clearance from live parts.
IPXXC	Protected against access with a tool. A tool ≥ 2.5mm diameter and 100mm long has adequate clearance from live parts.
IPXXD	Protected against access with a wire. A wire ≥ 1mm diameter and 100mm long has adequate clearance from live parts.

7671) often specify both types of code in regulations, for example an enclosure must comply with IP2X or IPXXB.

Table 9.2 summarises the additional codes. You will notice that they are very similar to the IP1X, IP2X, IP3X and IP4X codes defined above.

It is essential that inspectors and those being assessed have a good understanding of the IP codes and their application. Typical areas where IP codes are used in the *IET Wiring Regulations* (BS 7671) include: the requirements for 'Barriers and Enclosures' (regulation 416.2 on page 70) and, in Part 7, 'Special Installations or Locations', which starts on page 205. You should familiarise yourself with the IP codes used in these areas of BS 7671, as they often come up in exam questions.

SPECIAL INSTALLATIONS OR LOCATIONS

Part 7 of the *IET Wiring Regulations* (BS 7671) contains requirements for 'Special Installations or Locations'. Some types of installation and certain locations require additional regulations, above and beyond those specified in Parts 1 to 6 of BS 7671, to ensure their safety. Therefore, each section in Part 7 lists requirements for a particular 'Special Installation or Location'.

For example section 701 covers locations containing a bath or shower and section 710 covers medical locations. Those performing inspection and testing should have a good understanding of which types of installation and which locations are covered by Part 7. An inspector should also have a detailed knowledge of the regulations contained in the most common type of special installations and locations. Often those taking inspection and testing exams get caught out by questions relating to special installations and locations because they have not sufficiently revised in this area. One notable example was the inclusion of questions relating to a caravan park, which left many candidates stumped because they were not familiar with this type of installation in their day-to-day work.

FIGURE 9.2   Some locations require additional regulations to ensure their safety

CHAPTER 10

Testing an electrical installation

As I mentioned in the Introduction, electrical testing involves using electrical test equipment to take measurements. We do this to confirm the suitability of parts of the electrical installation that cannot simply be inspected using our senses. Electricity by its very nature is not much to look at; we cannot hear resistance or smell voltage, but these things are essential to the safety of the installation. Therefore, initial verification and periodic inspection and testing both rely on electrical testing as an integral part of their structure.

Electrical testing generally falls into two categories, **dead testing** and **live testing**. Dead tests, as the name suggests, are tests carried out on installations, circuits or equipment that have been safely isolated from the electricity supply, following the procedures detailed earlier in this book. Some tests however, cannot be performed on a 'dead' installation and therefore some degree of 'live' testing is always necessary. Naturally live testing introduces inherent hazards and should only be performed where the requirements of regulation 14 of the *Electricity at Work Regulations 1989* can be met. So before performing any live testing we must be sure that it is unreasonable to do this test any other way (i.e. dead test, calculation, inspection etc.), it is reasonable to perform this test 'live' and all precautions for safety have been taken to prevent danger or injury to the inspector and to others.

SEQUENCE OF TESTING

What tests are required, and what order they should be done in, really sits at the heart of understanding inspection and testing. Unfortunately this is another area where initial verification and periodic inspection and testing are handled differently. Below I have attempted to lay out the BS 7671 testing requirements.

Initial verification testing requirements

Regulation 612.1 on page 198 of the *IET Wiring Regulations* (BS 7671) explains the testing requirements with regards to initial verification. Regulation 612.1 tells us that we should carry out the tests listed in regulations 612.2 to 612.13 and compare their results with relevant criteria. Also, where relevant, the tests required by 612.2 to 612.7 shall be carried out in order and before the installation is energised. So regulation 612.1 reinforces the inspection and testing approach that we use with regards to new installations, i.e. inspection first followed by 'dead' testing and then 'live' testing, gradually step by step proving the safety of an installation that is new and has never worked before.

Remember that initial verification is a 100 per cent check of the complete installation, so we must always test the complete circuit and perform testing at all points on that circuit. Just testing up to a switch or only at the furthest point on a circuit is not acceptable. Regulation 612.1 also includes an instruction that if any test fails to comply, that test and any tests that may have been influenced by the failure are repeated again after the fault has been rectified.

Periodic testing requirements

The testing of existing installations differs in approach from that detailed above for initial verification. BS 7671 does not give much guidance in relation to periodic testing (621.2 on page 203) and simply suggests that inspection should be supplemented by appropriate tests from chapter 61, with the main aim of proving that the installation meets the disconnection times set out in chapter 41. The inspector therefore has some latitude to select only the tests necessary to make a decision with regards to the continuing safety of the installation. The use of sampling will normally play a big part during periodic testing and the rigid sequence of tests, adhered to for initial verification, will be applied in a more fragmented way. This is largely due to the fact that existing installations will usually be in service immediately prior to the start of the inspection and testing activity, making the step-by-step initial verification structure moot. The periodic testing sequence is based more on availability and access to circuits, which requires a lot of thought and planning.

The standard initial test sequence

Table 10.1 lists in order the standard tests as defined in chapter 61 of BS 7671 (starting on page 199).

Table 10.1   Standard tests as defined in chapter 61 of BS 7671

Test	Regulation
Continuity of conductors	612.2
Insulation resistance	612.3
Protection by SELV, PELV or by electrical separation	612.4
Insulation resistance/impedance of floors and walls	612.5
Polarity	612.6
Earth electrode resistance	612.7
Protection by automatic disconnection of supply	612.8
Earth fault loop impedance	612.9
Additional protection	612.10
Prospective fault current	612.11
Check of phase sequence	612.12
Functional testing	612.13
Verification of voltage drop	612.14

The tests above are applied only where relevant to the installation under test, for example the earth electrode resistance test is only applicable to those installations with an earth electrode. Later in this book I will address each test in turn.

TEST EQUIPMENT

To facilitate electrical testing the inspector will need suitable electrical test equipment, along with leads, probes and adaptors. Earlier in this book I looked at the safety requirements for test equipment detailed in documents such as *HSE Guidance Note* GS38. In this section I want to look in more detail at the general requirements relating to test equipment.

IET Guidance Note 3 details the requirements for test equipment in chapter 4, which starts on page 89. There are two main harmonised standards that relate to the test equipment used for inspection and testing, BS EN 61010, which details safety requirements, and BS EN 61557, which relates to the performance requirements. All modern test equipment should comply with these standards and it is important to confirm this in relation to any equipment that you are going to use.

Ideally, test equipment will be marked with the standard numbers and this will also be detailed in the user manual provided with the test equipment. To give you additional confidence of compliance, test equipment that has been independently tested by organisations such as BSI, UL or VDE will carry their certification mark. You should always be aware of the possibility of substandard or counterfeit test equipment, which may claim to be compliant with the appropriate standards, but which fails to comply and presents a potential danger to the user. Never use any test equipment that you believe may not comply with the appropriate standards.

Prior to use, all test equipment should be subjected to a thorough examination to ensure that it is fit for purpose and to identify any damage that may have been sustained since the test equipment was last used. The condition of any batteries should also be confirmed and new batteries fitted where required. Where possible the test equipment should be functionally tested. For example, in the case of a low resistance ohm meter, the two test leads can be connected together and resistance of the leads measured to prove functionality. An insulation resistance tester can also be tested by shorting the leads together. In addition to confirming that the equipment is safe to use and functions correctly, we must also be sure that the equipment is accurate.

FIGURE 10.1   It is essential to confirm that the equipment is accurate

I strongly recommend that any test equipment used to confirm the safety of an electrical installation should be calibrated at least annually or in accordance with the manufacturer's instructions. Calibration should be performed by an approved calibration laboratory, who will check the instrument against known standards and issue a calibration certificate detailing the test results. Following any suspected damage to the test equipment or repairs, it may be necessary to have the instrument re-calibrated to ensure that accuracy has not been affected. You must keep any calibration certificates safe as they may be required at a future date to prove the validity of your test results if challenged. When selecting a calibration company, care should be taken. Do not just focus your decision on price, but consider the quality and accuracy of the calibration process and, in this regard, laboratories accredited by the United Kingdom Accreditation Service (UKAS) are hard to beat.

To supplement the formal calibration carried out by a third-party laboratory, it is also now considered good practice to perform your own in-house checks to prove ongoing accuracy and identify any measurement errors that may occur in the period between formal calibrations. This is usually achieved by the use of a calibration 'check box', often purchased from the manufacturer of your test equipment. The 'check box' is constructed to allow a full range of measurements to be taken with the equipment under test and these results are then recorded and compared with previous results to identify any variation and therefore potential errors. Check box tests are usually performed monthly or in accordance with your quality policy. Where errors are suspected, the instrument should be withdrawn from service until such time as its accuracy can be formally verified by re-calibration.

Electrical Safety First have produced a best practice guide (*Best Practice Guide 7*). This guide is freely available from their website (www.electricalsafetyfirst.org.uk/electrical-professionals/best-practice-guides/) and contains guidance on the accuracy and consistency of test equipment for electrical installations. The contents of this guide have been agreed by the majority of industry bodies and should be considered essential reading for those studying this subject.

During inspection and testing exams it is very common for the candidates to be asked questions relating to test equipment, especially what equipment is required for which test. Candidates are also questioned about the measurement units for each test and sometimes even the range of expected values. To help you revise for this type of question, I have included this information in Table 10.2, which contains details of the typical test instruments, units and range of values used for the most common tests. This list is not exhaustive and only aims to give examples to aid revision for exam questions. Further details can be found in chapter 4 of *IET Guidance Note 3*.

Table 10.2 Typical test instruments, units and range of values used for the most common tests

Typical tests	Instrument used	Units	Typical range of values
Continuity of conductors	Low resistance ohm meter	Ohm Ω	0.1 to 10Ω
Insulation resistance	Insulation resistance tester	Mega ohms MΩ	0 to 999Ω
Earth electrode resistance	Earth electrode resistance tester	Ohm Ω	10 to 200Ω
Earth fault loop impedance	Earth fault loop impedance tester	Ohm Ω	0.01 to 200Ω
Prospective fault current	Prospective fault current tester	Kilo amps kA	0.5 to 16 kA
Check of phase sequence	Phase rotation instrument	N/A	N/A
Functional testing of RCDs	RCD tester	Milliseconds mS	5 to 50 mS

Nowadays, however, the idea of having a separate test instrument to perform each type of test is unlikely to be the case. Most inspectors now prefer multifunction devices that perform all the necessary tests in one instrument. There is a range of multifunction devices available from test equipment manufacturers, with additional features available based on price. New inspectors should give careful consideration to selecting the right instrument. Before purchasing any test equipment take the time to understand the additional features and benefits, so you know what is worth paying that little bit extra for and what you can do without. One example is the automatic RCD testing feature found on many mid-range or higher-end testers. This feature can save you a lot of time on-site and therefore most inspectors would say it is worth the additional money. My general recommendation is to buy the best tester you can afford, as you are unlikely to be troubled by having additional features, but the drawbacks of a cheap tester with limited functionality may only become apparent once you start testing more complex installations, such as those with three-phase circuits.

TEST RESULTS

During testing you will obtain test results, which are recorded on the appropriate inspection and testing forms. For initial verification it is important that these test results are verified against the relevant criteria, i.e. values provided by the designer of the installation or values obtained from BS 7671.

Often with new installations the designer will have to make assumptions about an installation's general characteristics because the installation may not even physically exist at the time of design, so it would not be possible to establish 100 per cent accurate figures until the supply is connected. It is therefore essential in these cases that any assumptions made by the designer are verified by measurement at the earliest possible opportunity.

Also in new installations there may be differences between the design and the final, 'as fitted' state of the installation; these differences could affect the ability of the installation to meet the requirements of BS 7671. So where the finished installation differs from the design, verification of test results against BS 7671 is essential.

Results obtained during periodic inspection and testing are notoriously difficult to verify. Older installations often bear little resemblance to their original designs, if available, due to additions and alterations over the years. These types of installation will also show some signs of deterioration, altering test results from those that would have been seen when the installation was new. Verification of periodic inspection and testing results is achieved by evaluating the ability of the installation to continue to operate in a safe manner and the continuing effectiveness of protective measures applied for the safety of the installation. Results are usually compared with values from BS 7671, but for much older installations some research may need to be carried out before compliance can be confirmed. For example, the protective devices used in older installations may still operate in the disconnection times required, but data on these devices will no longer be included in the current edition of BS 7671, so the manufacturer's data will need to be found before we can say if test results are acceptable.

CHAPTER 11

Testing the continuity of conductors (regulation 612.2)

Usually the first test we carry out on any electrical installation is to confirm that any conductors required for safety are continuous, i.e. they have a low electrical resistance, signifying that they are not broken or damaged and that they connect to the correct points in the installation. In an ADS installation, we typically verify the continuity of all protective conductors and all the conductors of ring final circuits.

The continuity of a conductor is verified by using a low resistance ohm meter. Readings obtained may be compared with the expected resistances calculated using the conductor length and the standard resistance values per metre from *IET Guidance Note 3* table B1 on page 128 (*IET On-Site Guide* table I1 on page 190) or values of resistance provided by the designer of the installation.

You must also be aware when performing these tests that parallel current paths, such as metal pipe work or structural steel work, can mask conductor faults making readings appear lower than they should be, or even making a conductor that is open circuit appear to be intact. Great care should be taken to understand, and where possible eliminate, the effects of parallel current paths on these readings.

TESTING THE CONTINUITY OF PROTECTIVE CONDUCTORS

Within an ADS installation there are a number of types of protective conductor that require verification:

- The earthing conductor
- The main protective bonding conductors
- The supplementary protective bonding conductors (if present)
- The circuit protective conductors.

These tests are classified as 'dead tests', therefore they must only be carried out with the complete installation or, where appropriate, the individual circuit under test, safely isolated from all sources of electrical energy, as covered earlier in this book. Many testers wrongly believe that it is safe to disconnect protective conductors while the installation is energised. This is not true and can be very dangerous. Protective conductors can carry significant leakage currents and therefore represent a potential electric shock risk if disconnected while the installation is energised. Also, the purpose of protective conductors is fault protection. If an installation is energised and a fault occurs while the protective conductors are disconnected for testing, the fault protection will not function correctly and this could have very serious consequences. For the reasons detailed above it is essential that great care is taken when disconnecting protective conductors for testing. Good practice dictates that each conductor is only disconnected for the shortest time possible and reconnected again immediately once the test is complete.

Once the complete installation is safely isolated and proven 'dead', it is normal to start by confirming the continuity of the **earthing conductor**. The earthing conductor is usually the largest protective conductor in the installation and connects the main earthing terminal (MET) to the supplier's means of earthing (TN systems) or the earth electrode (TT systems). Where possible, to avoid false readings due to parallel current paths, the earthing conductor should be disconnected at one end prior to testing. Using a low resistance ohm meter, a resistance measurement is made end-to-end on the conductor and the results obtained should be compared with the values expected for a conductor of the same cross-sectional area and length.

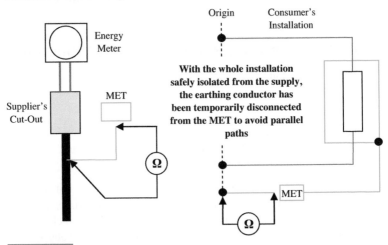

FIGURE 11.1   Diagrams showing typical measurement of earthing conductor continuity

Example

An earthing conductor with a cross-sectional area of 16 mm² and a length of 25 metres is measured. What is the expected resistance reading?

From table B1 (page 128 of *IET Guidance Note 3*), 16 mm² copper conductors have a resistance of 1.15 mΩ per metre at 20°C.

1.15 mΩ/m x 25 metres = 28.75 mΩ or 0.02875 Ω

Most low resistance ohm meters have a resolution of only 0.01 Ω and the accuracy of tests performed in the field is unlikely to be the same as that achieved in a laboratory. Therefore we would expect to see the actual measured resistance of the earthing conductor to be in the order of 0.03 Ω. Any values significantly higher than this should be investigated and may indicate a fault.

Due to the low values of resistance expected, it is important that the resistance of the test leads and probes does not influence the measured value. To avoid this, prior to testing we must take measures to avoid this error. Most low resistance ohm meters have a 'zero' or 'null' function, which allows lead resistance to be measured and then deducted automatically from future measurements. Where this function is not available, the tester may measure the lead resistance and then manually deduct this value from their future resistance readings. Often, when performing low resistance measurements on long conductors it will be necessary to use a long test lead (Wander lead) to make these end-to-end measurements possible. The resistance of these long test leads must also be zeroed/nulled prior to testing or measured and deducted manually after the test.

Following the verification of the earthing conductor, the tester can then proceed to verification of the **main protective bonding conductors**. Again, with the installation safely isolated, a low resistance measurement is made end-to-end on each main protective bonding conductor to confirm that the resistance is in the order of that expected by calculation using the method shown in the example above.

Parallel current paths are a particular problem when measuring main protective bonding conductors because of the necessary presence of extraneous conductive parts (e.g. water pipes, gas pipes or structural steel work), which can act as parallel current paths and therefore influence the test results. To ensure that results are not affected it is recommended that the main protective bonding conductor under test is disconnected at one end during this low resistance measurement.

To also verify the connection of the main protective bonding conductor to the extraneous conductive part, as part of this test I recommend that it is performed by disconnecting the conductor at the MET and then carrying out the measurement between the disconnected end of the conductor under test and the extraneous conductive part itself (e.g. to the water pipe or structural steel directly). This method allows us to ensure, for example, that there is a good low resistance electrical connection through the BS 951 pipe bonding clamp to the pipe work.

In the past I actually inspected one installation where the electrician had mistakenly bonded a plastic water pipe which was covered in paint. This had not been spotted during the initial verification because the tester had only tested to the end of the main protective bonding conductor and not to the pipe itself. So ever since I have always advocated testing to the extraneous part directly and not just to the end of the conductor or to the clamp.

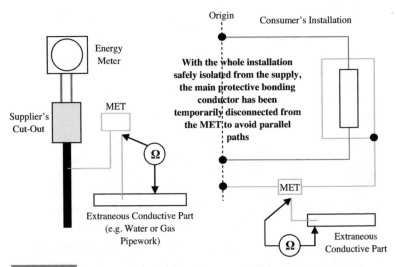

FIGURE 11.2   Diagrams showing typical measurement of main protective bonding conductor continuity

A common misconception relating to verifying the continuity of main protective bonding conductors is that this can be done visually if you can see the cable along its entire route. Unfortunately even the best electricians cannot 'see' resistance. We inspect the main protective bonding conductors as part of the inspection carried out before testing to confirm that they are present, adequately sized, correctly terminated and not damaged etc., but it is still necessary to also verify their continuity by a low resistance measurement.

In many installations, additional protection against electric shock is provided by the use of supplementary protective bonding. BS 7671 regulation 415.2 on page 69 of the *IET Wiring Regulations* relates to supplementary protective bonding and reminds us that in locations where this type of additional protection is applied, all simultaneously accessible exposed conductive parts of fixed equipment and extraneous conductive parts should be part of the supplementary protective bonding system. Also included in the system is a connection to the circuit protective conductors of equipment in the location and any socket outlets. The purpose of a supplementary protective bonding system is to ensure that all conductive parts in a location that can be touched at the same time are held at the same electrical potential as each other and as the general mass of earth. Therefore, a person touching any two conductive parts will experience no potential difference (voltage) across their body and so no shock current will flow. Also, by providing a connection between the supplementary protective bonding system and the general mass of earth we ensure that, if there is a fault between live conductors and any parts of the supplementary protective bonding system, automatic disconnection of supply will take place.

To verify a supplementary protective bonding system, we must carry out measurements using a low resistance ohm meter between all simultaneously accessible exposed conductive parts, extraneous conductive parts and circuit protective conductors within the location. The aim of these tests is to confirm that all parts that form the system are effectively connected together by a suitably low resistance (regulation 415.2.2).

For safety reasons these tests are usually performed with the installation safely isolated from the supply, and the test instrument must be zeroed or nulled prior to taking any resistance measurements for increased accuracy. As explained above, I would advocate making these

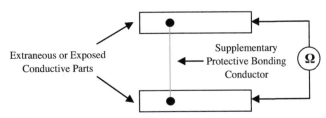

FIGURE 11.3 Diagram showing typical measurement of supplementary protective bonding conductor continuity

measurements between the parts themselves and not end-to-end on the individual conductors, because verifying the continuity of connections such as pipe bonding clamps is an important part of this process too. While parallel paths must be considered, we are concerned with the integrity of the system as a whole. So, therefore, in this context parallel paths are just part of the supplementary protective bonding system along with the supplementary protective bonding conductors.

CONTINUITY OF CIRCUIT PROTECTIVE CONDUCTORS

Regulation 411.3.1.1 on page 55 of the *IET Wiring Regulations* (BS 7671) reminds us that, in an ADS installation, a circuit protective conductor shall be run to and terminated at each point in wiring and at each accessory. This also includes points where the CPC is not currently necessary (i.e. a connection point for a Class 2 light fitting), because Class 1 equipment requiring an earth connection may be fitted in the future.

To ensure compliance with the above, and in addition to inspection, we must perform electrical testing to verify that all circuit protective conductors (CPCs) are electrically continuous and have a low electrical resistance. This resistance should be comparable with that expected by calculation using standard resistance values (*IET Guidance Note 3* table B1 on page 128) and the length of the circuit or values provided by the designer of the installation. There are two common methods of verifying the continuity of circuit protective conductors, referred to in *IET Guidance Note 3* as method 1 and method 2, but generally known by electricians as the $R_1 + R_2$ method and the R_2 or wander lead method. Each method has its own strengths and weaknesses and an experienced tester will use the method that best fits each situation and not stick religiously to one method. Below I will explain the two methods and their main differences. You can also read the descriptions contained in *IET Guidance Note 3*, starting on page 35. To help your understanding of the two methods, I find it easier to explain them in reverse order (i.e. method 2 first) so please bear with me.

To perform these tests in a safe manner, it is necessary for the installation (or where appropriate the individual circuit) to be safely isolated from the electrical supply. These tests are performed using a low resistance ohm meter and this meter should be zeroed or nulled prior to any tests being carried out and after any changes to the test lead or probe configuration.

The R_2 or wander lead (test method 2)

This test is performed by making resistance measurements between the Main Earthing Terminal (MET) and the earth terminals at each point in wiring and at each accessory. Because we are effectively measuring the resistance of the CPC, or more specifically that part of the earth fault loop defined as R_2 in BS 7671, this is known as the 'R_2 test'. This method will inevitably involve taking a large number of resistance measurements throughout the installation, often in positions distant from the MET. To facilitate these measurements we will require a long test lead, often referred to as a wander lead, which is connected to the MET. The tester will then use the other, shorter, test lead and probe to make contact with each earth terminal in turn and make resistance measurements. The highest resistance reading for each circuit is then recorded on the schedule of test results in the R_2 column.

The R_2 test method has the obvious advantage of being quick and simple to achieve and requires very little disruption to the installation, such as the disconnection of conductors or dismantling of equipment. There is however the trip hazard created by the long wander lead, which must be taken into account when testing occupied installations.

The R_2 method is particularly well suited to periodic testing due to its light touch approach and is ideal for use as part of a sampling strategy. The main down side for the R_2 test is that it only provides an R_2 value, which may seem to be sufficient. However, test method 1 provides a measured value of R_1 and R_2, which can then be added to the external earth fault loop impedance (Z_e) to calculate the circuit earth fault loop

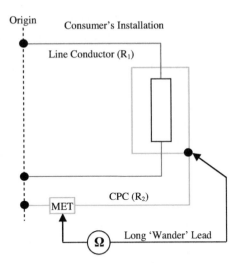

FIGURE 11.4   Diagram showing typical measurement of circuit protective conductor continuity by the R_2 method

impedance (Z_s) and therefore remove the need for a live measurement of Z_s and the risks associated with live testing. Where only R_2 measurements have been made, it will be necessary to also perform live earth fault loop impedance tests to confirm that the requirements for automatic disconnection have been met.

The R_1+R_2 test (test method 1)

This test method is ideal for initial verification testing, where conductors and accessories are readily accessible. One of the main drawbacks to the R_2 method is the need for a long test lead, which must run back to the distribution board and causes a trip hazard. Test method 1 removes the need for a separate long lead by using the circuit's line conductor (R_1) instead. As the name suggests the R_1+R_2 test measures the resistance of both the line and CPC conductors in one go.

For each circuit a temporary connection is made between the line and CPC conductors at the distribution board, and then a low resistance measurement is made between the line and earth terminals at each point in wiring and at each accessory. The highest of the readings obtained is recorded on the schedule of test results in the R_1+R_2 column. As mentioned above, the R_1+R_2 value can then be added to the external

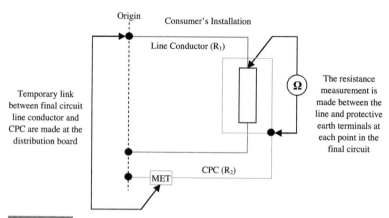

FIGURE 11.5   Diagram showing typical measurement of circuit protective conductor continuity by the R_1+R_2 method

earth fault loop impedance (Z_e) to calculate the circuit earth fault loop impedance (Z_s) and remove the need for a live measurement, making the overall testing process safer.

There are many ways to make the temporary connection between the line and CPC conductors at the distribution board. *IET Guidance Note 3* shows inserting a temporary link between the output side of the circuit protective device and the MET. I have reservations about this method because it has been known for these links to be left in by mistake, which is extremely dangerous if the installation is then re-energised. Also, the addition of any temporary links or additional connections can increase the resistance reading and decrease accuracy. I prefer to remove the line conductor from the output of the circuit protective device and connect it directly to a spare terminal in the MET. This method minimises connection resistance errors and ensures that if the connection is mistakenly left connected it will not result in a dangerous situation. If you have concerns about parallel current paths affecting the validity of your tests you may want to disconnect both the line and CPC conductors from the distribution board and connect them using a separate terminal block. Whichever method you choose to connect the line and CPC, always consider safety and accuracy when making your decision.

The R_1+R_2 test method is also preferred by many testers because it can help us identify some potential polarity problems too. BS 7671 tells us that single-pole protective and switching devices must be fitted in the line conductor only and that the centre contact of Edison screw lamp holders must be connected to the line conductor. By operating single-pole devices during the $R_1 + R_2$ test we can confirm that they do indeed break the line conductor and have not been accidentally fitted in the neutral by mistake. Also, we can test to the centre contact of Edison screw lamp holders confirming their correct connection to the line conductor.

Testing earthed metallic wiring systems

Although not very popular nowadays, it is still acceptable to use a metallic wiring system (for example metal conduit or SWA armour) as the circuit CPC. Many years ago we were required to confirm the integrity of this type of CPC by performing a high current resistance measurement using a specialist piece of test equipment designed for the job. In recent times the practice of high current testing has been removed from the guidance documents. We are currently advised that this type of system should instead be visually inspected along its length to confirm connections are sound. The continuity is then verified using a normal low resistance ohm meter by performing either test method 1 or test method 2, described above.

CONTINUITY OF RING FINAL CIRCUIT CONDUCTORS

Although commonly referred to as the 'ring circuit' or 'ring main', the correct title is currently 'ring final circuit'. So please be careful in exams or assessments to use the correct title or you may lose marks.

The ring final circuit is a unique type of circuit, because the design of its overload protection only works if the circuit is connected as a ring. Essentially, it relies on the continuity of the ring final circuit conductors for its safety. It is therefore important that we confirm this continuity as part of any inspection and testing activity. The safety of ring final circuits also requires that each point on the ring final circuit is correctly connected in terms of polarity and this must again be verified by testing. Finally, many inexperienced installers may wire ring final circuits incorrectly, especially when carrying out additions or alterations to the circuit. Often the ring final circuit may be bridged or connected in a figure of eight pattern and spurring from spurs is also a very common fault encountered.

To counter the above problems, and to verify that the circuit is in compliance with BS 7671, a special three-step test process has been created.

On successful completion of the three-step ring final circuit tests we will prove or measure the following:

- The ring final circuit conductors are all continuous.
- The ring final circuit is correctly wired as a ring.
- The polarity of each point on the ring is correct.
- We have an $R_1 + R_2$ value for the circuit.

The ring final circuit test is performed by measuring the resistance of the circuit conductors in various configurations, using a low resistance ohm meter. This is a dead test and may only be performed once the installation (or where appropriate the individual circuit) has been safely isolated from its supply. Also, as we are measuring low resistances, it is important to zero or null the meter before any measurements are made.

Step 1 – End-to-end measurements

The first step confirms the continuity of all the ring final circuit conductors. This is done by disconnecting all six cable ends from the distribution board and a resistance measurement is performed end-to-end on the line, neutral and circuit protective conductors. These readings

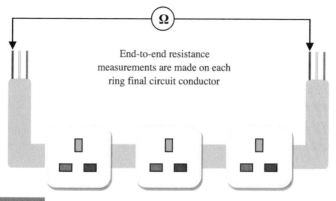

End-to-end resistance
measurements are made on each
ring final circuit conductor

FIGURE 11.6 Diagram showing typical measurement of ring final circuit conductors (step 1)

are recorded as r_1, r_n and r_2, respectively. Please note that a lower case 'r' is used to signify that these are end-to-end values to avoid confusion with $R_1 + R_2$, which use an upper case 'R'.

Once the three values have been measured, we can then initially verify that they are correct by comparing them with each other. In a ring final circuit where all three conductors are the same cross-sectional area, and also of similar length, the three readings should be approximately the same. We usually say that the readings should be within 0.05Ω as a rough guide. However, it is common for ring final circuits to have a reduced cross-sectional area conductor used for the CPC. For example, a $2.5mm^2/1.5mm^2$ cable is often used for ring final circuits and this cable has a smaller $1.5mm^2$ conductor used for the CPC. Where the CPC is smaller we would expect to see the r_2 value reading a higher resistance than that measured for the live conductors r_1 and r_n. In the case of a $2.5mm^2/1.5mm^2$, cable the measured resistance of the CPC should be approximately 1.67 times that of the values obtained for the live conductors.

The second method of verification requires some more detailed knowledge of the circuit, so it is more common during initial verification, where the information will have been provided by the designer. In this method we compare the measured values with calculated theoretical values either provided by the designer or calculated ourselves from the circuit length and cable data, such as that found in *IET Guidance Note 3* table B1 on page 128 (*IET On-Site Guide* table I1 on page 190).

Example

A ring final circuit wired using $2.5mm^2/1.5mm^2$ cable with a total length of 50m. From table B1 (*IET Guidance Note 3*, page 128):
   $2.5mm^2$ cable has resistivity at 20°C of $7.41m\Omega/m$
   $1.5mm^2$ cable has resistivity at 20°C of $12.10m\Omega/m$

So, if the measurements were taken at 20°C, which is normally assumed for inspection and testing purposes, we can calculate the expected values below.
   r_1 and r_n will be the same as they both have $2.5mm^2$ conductors
   So r_1 or $r_n = 7.41m\Omega/m \times 50$ metres $= 370.5m\Omega$

This is more commonly expressed as 0.37Ω, because most low resistance ohm meters display values in ohms to two decimal places.

And $r_2 = 12.10Ω/m \times 50$ metres $= 605mΩ$

Which again would usually be expressed as 0.60Ω.

These calculated values can then be compared with the measured values, and any significant differences between the two can be investigated. Please note that it is normal to see some discrepancy between theoretical and actual values due to many factors.

Successful completion of step 1 verifies that the ring final circuit conductors are not broken and that no bad connections are evident. We can now proceed to step 2.

Step 2 – Cross connection of line and neutral

To perform the second step of the three-step process we must now cross connect the line and neutral conductors at the distribution board and take measurements at each socket outlet or point on the ring final circuit. The key to getting this test right is correctly connecting the line and neutral conductors. First, you must identify the conductors at each end of the ring final circuit. You will have a line, neutral and CPC at end 1 and at end 2. Once identified, you may find it useful to use some tape to temporarily label these conductors.

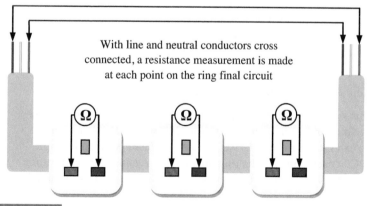

With line and neutral conductors cross connected, a resistance measurement is made at each point on the ring final circuit

FIGURE 11.7   Diagram showing typical measurement of ring final circuit conductors (step 2)

Using a terminal block, connect the line conductor of end 1 to the neutral conductor of end 2 and vice versa. We then perform a resistance measurement between line and neutral at each point on the ring final circuit. When performing measurements at socket outlets it will be necessary to use a suitable test adaptor.

To verify the readings obtained above it will be necessary to perform a quick calculation to establish the expected value. The calculation uses the r_1 and r_n values obtained in step 1.

The expected reading at each point on the ring final circuit will be approximately one quarter of the resistance of the line and neutral conductors added together.

For example using the circuit previously described above:

Expected value = $(r_1 + r_n) / 4$
Expected value = $(0.37 + 0.37) / 4$
Expected value = $0.74 / 4 = 0.185\Omega$

So, at each point on the ring final circuit we would expect to see a reading of about 0.185Ω. As mentioned earlier, low resistance ohm meters generally only measure to two decimal places and there will be some difference between theoretical and real-world measured values, so a reading in the order of 0.18Ω will be satisfactory. The important thing to verify in this step is that the values obtained at each point are the same, again usually within 0.05Ω as a rough guide. Any points reading significantly differently from the others should be investigated. Points that are spurred from the ring final circuit will naturally read higher and this must be taken into account. No values are recorded for step 2 but after successfully completing this step we can then move on to step 3.

Step 3 – Cross connection of line and CPC

This step is identical to step 2, however in this step we cross connect the line conductor of end 1 with the CPC of end 2 and vice versa. We can then perform a resistance measurement between line and CPC at each point on the ring final circuit. Again, when performing measurements at socket outlets it will be necessary to use a suitable test adaptor.

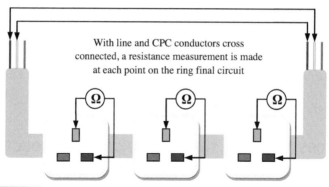

With line and CPC conductors cross connected, a resistance measurement is made at each point on the ring final circuit

FIGURE 11.8 Diagram showing typical measurement of ring final circuit conductors (step 3)

To verify the readings obtained above it will again be necessary to perform a quick calculation to establish the expected value. The calculation uses the r_1 and r_2 values from step 1.

The expected reading at each point on the ring final circuit will be approximately one quarter of the resistance of the line and CPC added together.

Again, using the circuit previously described above as an example:
Expected value = $(r_1 + r_2) / 4$
Expected value = $(0.37 + 0.60) / 4$
Expected value = $0.97 / 4 = 0.2425\Omega$

So we would expect to see a reading at each point on the ring final circuit of about 0.24Ω. As with step 2, the important thing to verify is that the reading at each point is roughly the same. Again, values that are not within about 0.05Ω of each other should be investigated as a potential fault. Any points wired as spurs will read higher than those points connected directly to the ring.

The highest reading obtained during step 3 also represents the $R_1 + R_2$ value for the ring final circuit, which is then recorded on the schedule of test results.

After all three steps have been successfully completed, care must be taken to remove any temporary connections and correctly reconnect the conductors at the distribution board.

By completing the three steps we have now established that the ring final circuit conductors are continuous and that the circuit is correctly wired as a ring. This test also allows us to confirm polarity for each point on the circuit. While it is normal to perform the ring final circuits test at the distribution board, we can use any point on the ring and it is often preferable to perform the test at a readily accessible socket outlet, if the distribution board is not easy to get to.

FACTORS AFFECTING THE RESISTANCE OF A CONDUCTOR

When carrying out inspection and testing it is important that we have a good understanding of the factors that affect our readings. Specifically, when performing the continuity of protective conductors test, we are concerned with the factors that affect resistance of conductors.

The resistance of a conductor is dependent on four factors. You can remember these by using the acronym **MALT**.

Material

Each material has a different atomic structure, therefore each material will have a different resistance per metre, usually expressed in mΩ/m; this property is known as its resistivity. Usually this property has a limited effect during inspection and testing as most conductors are made of standard materials, such as copper, and these materials are unlikely to change during the life of the installation.

Area (cross-sectional area)

The cross-sectional area of a conductor, when increased, allows electrons to flow more easily through it. This has the effect of lowering the conductor's resistance. The opposite effect can be observed when a conductor's cross-sectional area is reduced, thus increasing its resistance. The cross-sectional area of a conductor is therefore said to be inversely proportional to its resistance.

If the cross-sectional area of a conductor is doubled, its resistance halves. If we halve the cross-sectional area, the resistance will double.

This is often an area explored during exam questions, a typical example being:

You have a 25 mm² cable with a measured resistance of 0.8 ohms. If this cable were to be changed for a 50 mm² cable what would be the effect on the measured resistance?

If we double the cross-sectional area (CSA) we halve the resistance, therefore the answer would be 0.4 ohms.

In many cases, we see the effect of parallel resistance paths during inspection and testing, for example two cables connected in parallel. Where two cables are connected in parallel, this has the effect of increasing the cross-sectional area, i.e. two 2.5 mm² conductors connected in parallel would have the equivalent CSA of one 5 mm² conductor, so the resistance of the two conductors in parallel would be half that of either conductor on their own.

It is common to see the effect of parallel resistances when performing tests on main protective bonding conductors, where structural steelwork or metal pipes can introduce parallel paths and will reduce resistance readings. The effect of parallel paths can be to make resistance values appear lower, which could mask a defect. Great care should always be taken to consider the impact of parallel paths on your readings and reduce them where possible.

Length

The length of a conductor is often a key factor that must be considered during inspection and testing, specifically in relation to testing long protective conductors. If we increase the length of a conductor we increase its resistance, and if we shorten a conductor its resistance will decrease. Therefore the length of a conductor and its resistance are said to be directly proportional.

If the length of a conductor doubles then its resistance will also double; if the length is halved then again the resistance will halve.

This property is also explored during exam questions, a typical example being:

You have a 50 m protective conductor with the measured conductor resistance of 1 ohm. If the length of the lead is reduced to 25 m what will the new measured resistance be?

In this case halving the cable length will halve the resistance, so the correct answer would be 0.5 ohms.

Temperature

The temperature of a conductor also affects its resistance. If we increase the temperature of a conductor we will also increase its resistance; if we lower the temperature of a conductor we will decrease its resistance.

While performing inspection and testing we usually assume that testing takes place at a nominal ambient temperature of 20°C, however if a high level of accuracy is required the tester should take this into account and correction factors can be used to alter the values accordingly. Later in this book I will address earth fault loop impedance testing, where we use the 80 per cent rule to correct our maximum acceptable earth fault loop impedance values to allow for the increased resistance of our conductors at the cable's maximum operating temperature.

CHAPTER 12

Testing insulation resistance (regulation 612.3)

In most electrical installations, insulation of live parts is used as a method of basic protection and in some protective measures insulation is also used as a method of fault protection too. As basic protection, the insulation is there to stop hazardous currents from passing through the body of persons or livestock in normal, fault-free conditions. It is, therefore, essential that the effectiveness of the insulation is confirmed during all inspection and testing activities.

During initial verification, the main aim of the insulation resistance test is to confirm that the insulation has not been damaged during installation and that incorrect termination has not caused live parts to become connected to exposed or extraneous conductive parts by mistake. When performing the insulation resistance test as part of periodic inspection and testing, we already know that the installation has been in service for a period of time and that it should have passed the initial insulation resistance tests when it was new, so the emphasis during periodic testing is slightly different.

During periodic insulation resistance testing, we are looking for insulation faults that may have appeared during the life of the installation, such as water ingress or thermal damage. We are also trying to establish to what extent the insulation has deteriorated due to ageing. All insulation ages at different rates depending on the materials used, its environment and the operating temperature, so it is only by measurement that we can see if the installation is still safe for continuing service. We can also compare different sets of results from past inspection and testing activities with our current measurements to establish the rate of deterioration and use this information to inform our decisions about the length of time to the next periodic inspection and test.

Insulation, as the name suggests, is made from materials that are good insulators and therefore have a very high electrical resistance measured in mega ohms (MΩ). Common insulation materials include air, PVC, rubber or XLPE. We use the insulation resistance test to place the insulation under stress, in an attempt to simulate the electrical force that may be present during everyday use. Establishing the safety of the insulation within an installation is an essential part of inspection and testing, as any insulation faults can give rise to serious danger if the installation is energised. Therefore, the insulation resistance test is always performed as part of 'dead testing' with the complete installation (or where appropriate the individual circuit) safely isolated from its supply. The test voltage used, however, can be as high as 1000 VDC, so live working precautions should still be followed during these tests. Care must also be taken to ensure the safety of those present while these tests are carried out, as exposed metal parts of the installation could become live during testing and may represent a safety hazard.

The aim of these tests is to confirm that all the insulation within the installation has a high enough electrical resistance to effectively separate all live parts from earth and from all other live parts. A recent change to BS 7671 now requires that when performing any insulation resistance tests to earthed metal parts (for example the armouring of an SWA cable), we ensure that these parts are connected to the earthing arrangement (MET) during the test. This is an added safety precaution against electric shock during testing and also allows us to detect general earth faults (for example, a short circuit to metal pipework) that may have previously escaped our notice. Those who learned inspection and testing before the introduction of this new method must ensure that they now adopt this new method.

When testing the insulation within an installation, the preferred method is to test the complete installation in one go, by testing at the distribution board with the complete installation safely isolated from its supply. This method is usually adopted for initial verification and is simple to carry out on new installations. In the case of periodic inspection and testing, fault finding or maybe additions and alterations, it may be appropriate to test each circuit individually. This is acceptable but may be much slower.

Like decorating your house, the key to insulation resistance testing is preparation, preparation, preparation. If the installation is correctly prepared, the actual tests take very little time at all, but failing to set up the installation ready for testing can lead to false readings, missed defects and even damaged equipment and appliances. So always take your time to set up the installation before insulation testing.

Essentially there are two quite similar things that we need to consider before an insulation resistance test can be carried out. These are 'precautions' and 'preconditions'. Precautions are measures taken to protect the installation during the test and also those present during testing as well. Preconditions are used to set up the installation to ensure that the test is valid and we get meaningful results.

Typical precautions (not exhaustive) include:

- The installation is safely isolated from its supply.
- For three-phase TN installations with only three-pole isolation, the neutral must also be safely isolated, usually by removal of the neutral link.
- All live working precautions are applied in-line with your company policy and risk assessments.
- All sensitive electronic equipment must be disconnected to avoid damage caused by the test voltage.
- Any capacitors or storage batteries are disconnected.
- All equipment covers and enclosures are in place.
- No person is exposed to parts that could become live during testing.

Typical preconditions (not exhaustive) include:

- All fuses in place, switches and circuit breakers closed (excluding those used for safe isolation of the installation/circuit above).
- Disconnect all current using equipment.
- Connection of earthed parts under test to the earthing arrangement (MET) within the installation.
- Two-way and intermediate switch for lighting should be tested in all switch positions.
- Where contactors are used, measures must be taken to ensure that any wiring on the load side of the contactor is not excluded from insulation testing. This may involve temporary link connections or alternative testing on the load side wiring.
- Any dimmer switches or neon indicators may need to be removed and the connections temporarily linked during testing.

Only when we are sure we have fully considered the precautions and preconditions applicable to the installation under test and all these measures have been implemented, can we then proceed to insulation testing. To perform the insulation resistance tests, we use an insulation resistance tester, which measures the resistance of the insulation by placing a high DC voltage across the insulation. Depending on the nominal voltage of the circuit under test, we must select a suitable DC test voltage.

The insulation resistance tester expresses values of insulation resistance in mega ohms (MΩ). These values may range from 0.00 MΩ indicating a possible short circuit, up to the full range of the tester, often 999 MΩ. Ideally, we would like to see this very high insulation resistance reading for all tests, especially when testing new installations, but for older installations the expected results may be lower. BS 7671 includes table 61 on page 199, which details applicable DC test voltages and minimum acceptable insulation resistance values. I have summarised this information in Table 12.1.

Before performing any insulation resistance tests it is important that we set the insulation resistance tester to the correct test voltage and that we confirm correct operation of the tester and any associated leads or adaptors by shorting the probes together and performing a test. The insulation resistance tester should read 0.00 MΩ, correctly identifying the short circuit we created by connecting the probes. Failure to perform this simple test could lead to false readings being obtained, if for example an internal fuse was blown inside the tester or a test lead was open circuit. In these cases, all readings would appear OK irrespective of the state of the insulation.

As mentioned above, the insulation resistance test is usually performed at the distribution board, where tests are carried out between live conductors and between live conductors and the earthing arrangement.

Table 12.1 Applicable DC test voltages and minimum acceptable insulation resistance values

Circuit	Test voltage (DC)	Minimum insulation resistance
SELV or PELV	250V	0.5MΩ
≤ 500V (Not SELV or PELV)	500V	1MΩ
> 500V	1000V	1MΩ

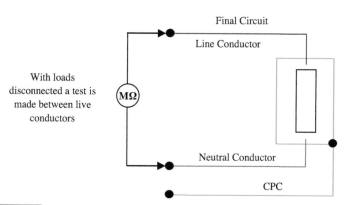

FIGURE 12.1 Diagram showing typical measurement of insulation resistance between live conductors

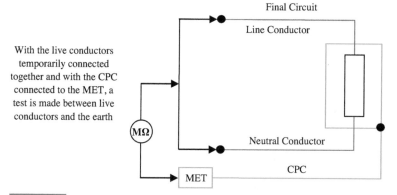

FIGURE 12.2 Diagram showing typical measurement of insulation resistance between live conductors and earth

For a single-phase installation, the three individual tests would be:

- Line to Neutral
- Line to Earth
- Neutral to Earth.

For a three-phase installation, the ten individual tests would be:

- Line 1 to Line 2
- Line 1 to Line 3
- Line 2 to Line 3

- Line 1 to Neutral
- Line 2 to Neutral
- Line 3 to Neutral
- Line 1 to Earth
- Line 2 to Earth
- Line 3 to Earth
- Neutral to Earth.

In practice, many of the individual tests listed above can be combined (regulation 612.3.1) by connecting the line and neutral conductors together. For example, the ten three-phase tests above could be reduced to just five tests using this principle: L1 to L2, L1 to L3, L2 to L3, L1-L2-L3 connected together to neutral and L1-L2-L3-neutral connected together to earth. As you become more experienced in inspection and testing you will be able to use combining to increase the speed of your testing, but for those new to the subject you may find it easier to stick to individual tests to start with. Remember you may also need to perform additional tests where, for example, two-way switching is used to full test the complete installation.

Regulation 612.3.3 allows that where a circuit contains electronic devices that may influence the test results or be damaged, we may test only between live conductors connected together and earth, omitting the test between live conductors. However, regulation 612.3.2 suggests that our first approach would be to disconnect any sensitive electronic equipment before testing and, if it is not reasonably practicable to remove this equipment, then the test voltage could be reduced to 250V DC to avoid damage, although the minimum value of insulation resistance must remain at 1MΩ. In the case of sensitive electronic equipment that cannot be disconnected during testing, the tester must decide which approach to adopt. Be warned that performing the insulation resistance test on this type of equipment without suitable precautions can cause serious equipment damage.

Once we have completed all the necessary measurements, the lowest values obtained between live conductors and between live conductors and earth are recorded on the schedule of test results. Where the complete installation was tested at the distribution board, these values can be entered for all the circuits tested.

Just as important as the precautions and preconditions before and during testing, is restoring the installation to its original state after testing.

All temporary connections should be removed and all disconnected equipment must be correctly reconnected to the affected circuits. Spend time checking each circuit to ensure nothing has been missed before moving on to the next set of tests.

FACTORS AFFECTING THE RESISTANCE OF INSULATION

When carrying out insulation resistance testing we must have a good idea in advance of what results to expect and we must know how to interpret the actual results obtained. During exams, examiners often aim to test your understanding in this area, so it is important that we spend a little time looking at the two main factors that affect insulation resistance that come up in most exams. These two factors are cable length and the number of cables connected in parallel. This subject is traditionally tricky to understand because it is generally the opposite from the way most people would guess.

Cable length

When cable length increases we learned earlier that the conductor resistance increases. However, in the case of insulation resistance, increasing the cable length has the opposite effect. A longer cable will have a lower insulation resistance. By increasing the length of the cable we are effectively increasing the CSA of the insulation and therefore the insulation resistance actually goes down.

Cables connected in parallel

When we connect additional cables in parallel, for example when we measure the insulation resistance at the distribution board feeding ten circuits, the effect is to increase the CSA of the insulation by ten times and this therefore reduces the overall insulation resistance. In practice the insulation resistance is still usually above the maximum range of the tester and therefore we rarely notice this effect in our measurements, but in older installations or those with damaged insulation this effect may be observed. In cases where we are testing circuits in parallel and low insulation resistance is discovered, it is recommended to test the circuits again individually to establish whether the low reading is caused by the combined effect of circuits in parallel or one faulty circuit dragging the reading down.

You are often asked in an exam to calculate the effects of parallel insulation resistances, so I have included an example below.

Example 1

We have five circuits with insulation resistances of 50 MΩ, 20 MΩ, 25 MΩ, 40 MΩ and 50 MΩ, respectively. What will be the insulation resistance reading when these five circuits are tested in parallel?

To answer this question we need to use the standard formula for calculating parallel resistances.

$$\frac{1}{Rt} = \frac{1}{R1} + \frac{1}{R2} + \frac{1}{R3} + \frac{1}{R4} + \frac{1}{R5}$$

So add in the numbers, keeping all values in MΩ.

$$\frac{1}{Rt} = \frac{1}{50} + \frac{1}{20} + \frac{1}{25} + \frac{1}{40} + \frac{1}{50}$$

Now you could just simply enter the above into a calculator, i.e. 1 divided by 50 equals, then add the values together. But modern scientific calculators can make this much easier. They will have a button labeled X^{-1} or $1/X$. This button allows us to type the above equation directly into the calculator in one go (i.e. 50 X^{-1} + 20 X^{-1} + 25 X^{-1} ... etc.). Performing the calculation in this way allows us to be very quick and accurate, so you should spend time mastering this function on your calculator as it will improve your exam performance.

So after calculation on the calculator:

$$\frac{1}{Rt} = 0.155$$

But we are not done yet, because we need Rt. So we must remember to do one final step.

$$Rt = \frac{1}{0.155} = 6.45\,M\Omega$$

And the final answer is that if we tested all five circuits while connected in parallel, we would see a reading of 6.45 MΩ, even though the lowest single reading is 20 MΩ. It is therefore clear that great care must go into interpreting the results obtained during insulation resistance tests, as a low reading does not always mean a faulty circuit.

Example 2

To catch out the less experienced candidates, examiners sometimes include questions like the one below.

You have a distribution board with 16 final circuits connected to it. Each circuit has an insulation resistance of 100 MΩ. What will be the insulation resistance of the complete installation when all 16 circuits are tested together?

Now the approach of the less experienced candidate will be to attempt the long calculation below.

$$\frac{1}{Rt} = \frac{1}{100} + \frac{1}{100} + \frac{1}{100} + \frac{1}{100} + \frac{1}{100} + \frac{1}{100} + \frac{1}{100} + \frac{1}{100} + \frac{1}{100}$$
$$+ \frac{1}{100} + \frac{1}{100} + \frac{1}{100} + \frac{1}{100} + \frac{1}{100} + \frac{1}{100} + \frac{1}{100}$$

This will get the job done, but will eat up valuable exam time and shows limited understanding. Instead, we can simplify the equation as shown below:

$$\frac{1}{Rt} = \frac{16}{100} = 0.16$$

And the final step:

$$Rt = \frac{1}{0.16} = 6.25\,M\Omega$$

As part of your revision, practise the types of question seen in examples 1 and 2 using your own values, until you become fluent in using your calculator and simplifying where possible to streamline your calculations.

CHAPTER 13

Testing polarity (regulation 612.6)

The term 'polarity' essentially means that we have got our wires the right way round, i.e. the lines in the line terminals, the neutrals in the neutral terminals and the protective conductors in the 'earth' terminals. Sounds obvious, but failure to observe correct polarity has caused many serious incidents in the past. Often polarity faults may go unnoticed for years only to surface under faulty conditions where they can prove fatal. So confirming correct polarity within the installation is a critical part of any inspection and testing activity.

Confirming correct polarity is a combined process of visual inspection, 'dead' testing and 'live' testing. There is the possibility for a polarity fault at every termination within the installation and even within the supplier's installation itself, so we need to be methodical in our approach to ensure that no polarity faults are missed.

The starting point for confirmation of correct polarity is our visual inspection. During the initial verification of new installations, 100 per cent of terminations are visually inspected to ensure cables are correctly identified by colour or number and that each conductor is connected to its correct terminal. During periodic inspection, sampling is used, so the inspector should concentrate on areas within the installation where polarity faults may have been introduced, such as additions, alterations or repairs. Specific attention should be paid to any DIY electrical work or work performed by unqualified electricians.

To supplement the visual inspection performed above, we gain additional confirmation of correct polarity during the performance of the 'dead' tests. These tests are not usually performed as a separate 'polarity test' but we use the information gained during other tests to confirm our visual inspection findings. The tests performed to confirm continuity of protective conductors and the continuity of ring final circuit conductors are very useful in this regard. While these tests do not 100 per cent prove correct polarity in their own right, they will highlight common polarity faults and back up our visual inspection. There are several typical polarity issues that give rise to danger and as a priority we must confirm these during dead testing: single poles protective and switching devices are connected in the line conductors only; correct termination of socket outlets; and the centre contact of Edison-screw lamp holders are connected to the line conductor.

CONFIRMATION OF SUPPLY POLARITY

Polarity faults in the supply to an installation are not unheard of and can be extremely dangerous if not detected and rectified immediately. It is, therefore, very important that during any inspection and testing activities we confirm supply polarity before the installation is energised. Confirmation of supply polarity is also an important part of periodic inspection and testing to identify any historic supply polarity faults that may have gone unnoticed.

As expected with any work carried out at the origin of the insulation, safety during the polarity test is paramount. Any measurements carried out on the supply side of the main distribution board are particularly dangerous due to the potential for very high overvoltages that can occur in this CAT IV environment. Please ensure that your live working risk assessment for this type of working includes the use of suitably rated PPE and CAT IV test equipment.

To verify the supply polarity, a series of measurements are made with an approved voltage indicator to confirm that the correct voltages are present at the correct supply terminals. As these measurements are generally carried out before the installation is energised, this test is usually performed on the incoming supply terminals of the main switch and earth connections are made to the MET.

For a single-phase installation, polarity tests are performed as in Table 13.1.

Table 13.1   Polarity tests for a single-phase installation

Test between	Expected voltage AC
Line and Neutral	230V
Line and Earth	230V
Neutral and Earth	0V

Table 13.2	Polarity tests for a three-phase installation
Test between	**Expected voltage AC**
L1 and Neutral	230V
L2 and Neutral	230V
L3 and Neutral	230V
L1 and Earth	230V
L2 and Earth	230V
L3 and Earth	230V
L1 and L2	400V
L1 and L3	400V
L2 and L3	400V
Neutral and Earth	0V

For a three-phase installation, polarity tests are performed as in Table 13.2.

By confirming that the readings taken match those expected, we can say that the supply polarity is likely to be correct. However these tests are not a 100 per cent guarantee. Especially in TN-C-S supplies, it is possible for some polarity faults to still produce the results in the tables above. Only polarity tests that use the general mass of earth as a reference will identify this type of fault; in the past it was common to do this using a neon screwdriver. Needless to say, the practice of using a neon screwdriver is now seen as hugely dangerous, so the modern equivalent would be a CAT IV rated noncontact voltage detector (volts stick) or a CAT IV single-pole contact type voltage detector. Using these devices to identify the line conductor(s) is only an additional test and should not be performed on its own as the sole proof of correct polarity. The schedule of test results requires that correct polarity is also confirmed at each distribution board.

Where there is still doubt with regards to polarity after visual inspection and 'dead' testing have been carried out, the tests listed above for confirming correct supply polarity can also be used to perform a live polarity test at points within the installation once they have been energised. Once correct polarity has been established for the supply and all points within the installation we can then proceed to the next test in the sequence.

CHAPTER 14

Testing earth electrode resistance (regulation 612.7)

Many electrical installations contain earth electrodes, which are relied on to make a good connection between the installation's earthing system and the general mass of earth. The most common use of earth electrodes is in TT type earthing systems, but they are also seen as part of lightning protection systems or used with generators, for example. The purpose of the earth electrode resistance test is to measure the resistance between the earth electrode and the general mass of earth and compare the measure value with the maximum values specified for each application. Earth electrode resistance can vary due to soil drying and freezing, so ideally testing would be carried out under least favourable conditions. In practice, however, this is rarely possible so the tester must consider the effect of soil condition on any reading obtained.

IET Guidance Note 3 details the three common earth electrode resistance test methods in chapter 2.6.14, which starts on page 52. The three methods are named E1, E2 and E3. Please take some time to study the descriptions in *IET Guidance Note 3* as these tests can be the subject of exam questions. This is especially true for test E1 where in the past candidates have been asked to describe this test with the aid of a diagram under exam conditions. Below I have summarised the three methods.

Note: For methods E1 and E3 it is necessary to disconnect the electrode from the installation's earthing system during testing. Never disconnect the earth electrode from a live installation. Installations must be safely isolated from their supply before the electrode can be disconnected. The disconnection must only be for the shortest time possible to allow the tests to be performed and reconnection must take place immediately following completion of the tests.

TEST E1

Test E1 requires the use of a dedicated earth electrode resistance tester and involves the use of temporary test spikes (electrodes) inserted into the ground to measure the resistance of the electrode under test. This test can be performed on a single earth electrode and does not require the presence of a mains supply, so it is ideal for testing the earth electrodes for generators. This test relies on two test spikes, the current spike and the potential spike. The use of temporary test spikes, however, can be problematic in built-up areas where concrete or tarmac surfaces restrict access to the general mass of earth.

The current spike is inserted into the ground at a suitably long distance from the electrode under test (*IET Guidance Note 3* suggests at least ten times the length of the electrode under test away). The potential spike is then inserted halfway between the other two spikes and a measurement is made by connecting all three spikes to the terminals of an earth electrode tester. Two further measurements are made with the potential spike moved to a position 10 per cent closer to the electrode under test and then 10 per cent further away. The three values obtained are averaged and then the percentage deviation of each reading from the average is calculated. Where the values deviate by 5 per cent or less the highest of the three figures can be recorded as the resistance of the electrode under test, in ohms. Deviation of more than 5 per cent is unacceptable and in this case the tests should be repeated with the current spike inserted further from the electrode under test.

TEST E2

Test E2 is a relatively new addition to the list and has been included due to advances in test equipment technology, which have led to changes in common testing practice and the widespread adoption of this test method for certain applications. Method E2 utilises a new specially designed clamp type tester, which has the major benefit of not requiring the electrode under test to be disconnected from the installation during the testing. There are many different clamp type testers becoming available on the market and their use and applications are all subtly different, so careful consideration of the manufacturer's instructions is important if you are considering this type of test method.

IET Guidance Note 3 mentions two of the most common types of clamp type tester: the single coil and dual coil type clamps. The single coil clamp test is very similar to the procedure for test E1, in that it uses temporary test electrodes to obtain its reading. The dual coil type clamp requires no test electrodes, but can only be used in systems with multiple earth electrodes, such are lightning protection systems or substation transformers.

TEST E3

Test E3 is used solely on TT earthing systems that are connected to the electricity supply network. Method E3 measures the resistance of the installation earth electrode by performing an earth fault loop impedance test at the origin of the installation, with the installation earth electrode disconnected from the rest of the earthing system to avoid parallel paths. This 'live' test is performed using an earth fault loop impedance tester and gives a value in ohms for the resistance of the earth fault loop. While this test is not a true measurement of the resistance of the installation earth electrode alone, it does give us a maximum resistance value for the earth fault loop, which includes the electrode under test. It would be true to say that $Z_e > R_a$ but it is impossible using this test method to place an exact value on the resistance of the installation earth electrode alone. So in the case of TT systems measured by using method E3 we allow the simplification that $Z_e = R_a$ (i.e. we say that the external earth fault loop impedance is the same as the electrode resistance).

EXPECTED TEST RESULTS

When testing an earth electrode we are trying to establish that it has a reliable connection with the general mass of earth and that this connection will remain stable all year round. In regulation 411.5.3, on page 60 of the *IET Wiring Regulations* (BS 7671), we are reminded that the resistance of an installation earth electrode should be as low as practicable and that values exceeding 200Ω may not be stable. So for TT installations this 200Ω figure is used as the benchmark for confirming an earth electrode is stable. If you are testing earth electrodes associated with other applications, such as generators or lightning protection systems, it will be necessary to consult the system's designer for other installations to assess compliance.

In TT systems protected by RCDs we must also consider the requirements for protection against electric shock laid out on page 60 of the *IET Wiring Regulations* (BS 7671). The maximum Z_s values in table 41.5 can mean that the resistance of the installation earth electrode needs to be lower in order for the final circuits to pass the Z_s requirements.

The maximum Z_s values are calculated using the formula below:

$$\text{Zs Max} = \frac{50}{I\Delta n}$$

So for a final circuit protected by a 30mA RCD ($I\Delta n = 0.03A$) the equation would be:

$$\text{Zs Max} = \frac{50}{0.03} = 1667\Omega$$

A value of Z_s this high is unlikely to place further restrictions on the resistance an earth electrode that already meets the 200Ω stability requirement. But if the RCD were rated at 300mA or 500mA the maximum Z_s values then would fall to 167Ω and 100Ω respectively and would therefore require a lower acceptance value for the resistance of the installation earth electrode. It is for this reason that many installers use a rule of thumb figure of 100Ω as the maximum acceptable value for all electrodes in RCD protected TT systems.

The connection between an earth and the general mass of earth deteriorate over a period of time as soil conditions change and due to corrosion of the electrode itself. It is therefore essential that all earth electrodes are subject to regular inspection and testing, as faults are unlikely to be evident to the user.

CHAPTER 15

Protection by automatic disconnection of supply (regulation 612.8)

Although listed in the sequence of tests, protection by ADS is not actually a test. The correct operation of the ADS system is verified by the successful completion of a range of other tests, for example the earth fault loop impedance tests. The results obtained are then compared with maximum values to confirm that the disconnection times specified in the *IET Wiring Regulations* (BS 7671) for ADS installations will be met.

EARTHING SYSTEMS

In the UK, the supply provided to the electrical installation generally falls into one of three types; these types are differentiated by the method of supplying a connection to the general mass of earth and therefore are commonly referred to as earthing systems. The three systems are named TN-S, TN-C-S and TT, and the names help to describe how earthing is achieved within the system, but first we need to know what each letter means.

Using the key in Table 15.1, we can now start to understand the meaning of the earthing system names.

Table 15.1 Earthing system letter codes

Letter	Definition
T	**T**erra (Latin for Earth)
N	**N**eutral
C	**C**ombined
S	**S**eparate

Table 15.2	Earthing system definitions
System	**Definition**
TN-S	In this system **T**erra (Earth) and **N**eutral are **S**eparate conductors.
TN-C-S	In this system the **T**erra (Earth) and **N**eutral are **C**ombined into one conductor in the supply and **S**eparated in the installation.
TT	In this system we use one earth electrode at the supply transformer to make the connection to Earth (**T**erra) and a second earth electrode to connect the installation earthing system to Earth (**T**erra).

TN-S EARTHING SYSTEMS

TN-S systems are perhaps the most straightforward to understand because, like most circuits, they use separate conductors for neutral and protective earth.

A majority of TN-S systems are now quite old and they can often be identified by their lead-sheathed supply cables. The lead sheath acts as the protective conductor and the line and neutral conductor are enclosed within it. This type of earthing system has a soldered joint connecting the lead sheath of the supply cable to the earthing conductor and often the earthing conductor has a small CSA and is green in colour. The soldered connection is a particular weak point and in many cases this connection may be broken. In the past, amateur or DIY electricians have attempted

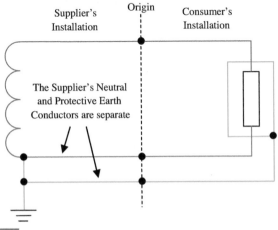

FIGURE 15.1   Diagram of a TN-S earthing system

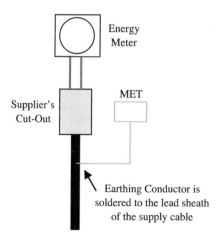

FIGURE 15.2 Identifying a TN-S earthing system

to remake this type of connection with a BS 951 pipe bonding clamp fastened around the lead supply cable. This practice is highly dangerous as it crushes the supply cable and can lead to a catastrophic failure. Where damage is suspected or you discover that a bonding clamp has been fitted in the past, this should be reported immediately to the DNO.

TN-C-S EARTHING SYSTEMS

This earthing system is the most common type of system used in the UK and is an improvement over the TN-S system because it does not use separate neutral and protective conductors in the supply. This allows the cables to be smaller, cheaper and easier to install. This type of system usually employs an insulated, concentric type supply cable. This type of cable has the line conductor(s) in the centre and concentric conductor wrapped around it, in much the same way as the coaxial cable used for TV aerials.

The concentric conductor is made up of many individual strands and this conductor acts as the Combined Neutral and Earth conductor (CNE), also known as the Protective Earth and Neutral (PEN) conductor. At the origin of the installation, inside the service cutout, a concealed terminal block allows the neutral and protective conductors to be separated. The TN-C-S earthing system can be identified by spotting the earthing conductor coming out of the side/top of the service cutout.

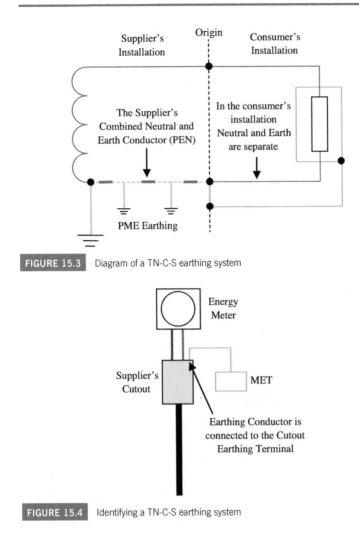

FIGURE 15.3　Diagram of a TN-C-S earthing system

FIGURE 15.4　Identifying a TN-C-S earthing system

There are different methods that can be used by the supplier to achieve a TN-C-S earthing system – the most common is a method call PME or Protective Multiple Earth. In addition to the earth electrode at the supply transformer, this method uses multiple earth electrodes along the length of the supply cable to connect the PEN conductor to earth. The use of multiple earth electrodes improves the resilience of the system and reduces the chance of losing the earth connection to the installation.

TT EARTHING SYSTEMS

The TT system is different from the TN systems because it does not have a metallic earth return path in the supply. Instead, the earth return path in the TT system is through the ground.

Like the other systems, there is an earth electrode at the supply transformer, but the TT system uses a second earth electrode within

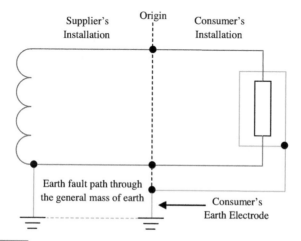

FIGURE 15.5 Diagram of a TT earthing system

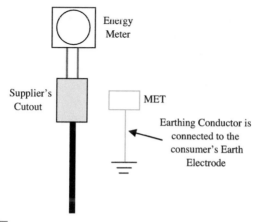

FIGURE 15.6 Identifying a TT earthing system

the consumer's installation to connect the earthing system directly to the general mass of earth. TT systems are often seen in rural locations, where originally it was not feasible for the supplier to use expensive lead-sheathed cable or PME systems. The TT system offered a quick, cheap and simple solution for installations that were spread out and often supplied by overhead LV cables.

The main advantage of the TT system is that it is independent of the supplier's installation and under the control of the consumer. TN systems rely on the integrity of the earth provided by the supplier and if there is a problem with this earthing, all you can do is call the DNO. In some cases TN systems are even converted to TT if there is a problem with the supply earthing. Because the installation earth electrode is the responsibility of the consumer, regular inspection and testing is essential. Many consumers are not aware that they have an earth electrode and it is common to find electrodes that have been concreted over or earthing conductors that have been cut off by mistake. A major disadvantage of the TT system is the comparatively high impedance of the earth path, which can be as high as 200Ω compared with TN systems, which are $\leq 0.8\Omega$ (TN-S) or $\leq 0.35\Omega$ (TN-C-S). The effect of this high impedance is to reduce the prospective earth fault current to a value below which normal overcurrent devices will operate.

$$\text{Ipefc} = \frac{\text{Uo}}{\text{Zs}} = \frac{230V}{200\Omega} = 1.15A$$

As you can see from the calculation above, the prospective earth fault current in TT systems can be as low as 1.15A, so overcurrent protective devices cannot be used for earth fault protection. In TT systems earth fault protection is provided by the use of residual current devices (RCDs), which will operate at much lower currents.

CHAPTER 16

Testing earth fault loop impedance (regulation 612.9)

Measurement of actual disconnection time is not really practical. Fun though it might sound to hammer a nail through the cable and try to time the disconnection with a stop watch, common sense tells us that this is not at all safe or accurate, so a better approach is required. This is why measuring the impedance of the earth fault path is so essential, because if we know the impedance and the voltage to earth U_o we can then calculate the prospective earth fault current. Then by using the graphs from appendix 3 of the *IET Wiring Regulations* (BS 7671) or the manufacturer's data we can check the theoretical disconnection time for the circuit and confirm compliance.

Figure 16.1 illustrates a typical earth fault path.

The earth fault loop impedance test is performed using an earth fault loop impedance tester, which gives values in ohms. This is a 'live'

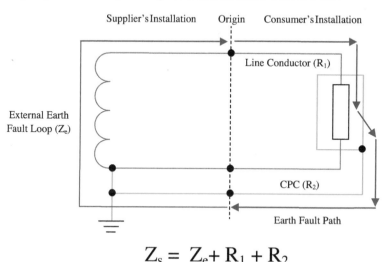

Supplier's Installation   Origin   Consumer's Installation

Line Conductor (R_1)

External Earth Fault Loop (Z_e)

CPC (R_2)

Earth Fault Path

$$Z_s = Z_e + R_1 + R_2$$

FIGURE 16.1   Diagram of a typical earth fault path (TN-S)

test performed with the circuit connected to its supply, so great care must be taken to fully risk assess the activity and apply all necessary control measures. In many cases it may be possible to minimise the amount of 'live' testing we do by using calculated values of earth fault loop impedance instead of measured values. For example, if we have a measured value of external earth fault loop impedance Z_e and a measured $R_1 + R_2$ for the circuit, we can simply add the two values to get the circuit earth fault loop impedance Z_s.

Calculation therefore removes the need for a 'live' test and its associated dangers. This is now seen as good practice as we try to minimise the amount of work carried out on or near 'live' equipment. A 'live' earth fault loop impedance test is now only performed where there is no other way to achieve the same results and all reasonable and practicable precautions have been taken.

Most common earth fault loop impedance testers offer two types of tests: **high current** and '**no trip**'. These tests may also be further broken down into three-lead or two-lead options. The high current test is the most accurate and will usually only require the use of two test leads (Line and Earth). The downside of this test is the fact that it is not suitable for circuits protected by an RCD. Because a high current flows between line and earth during the test, any RCDs will see this as a fault and immediately disconnect the circuit.

In years gone by electricians were encouraged to link out the RCD during these tests to avoid unwanted tripping, but this practice is now seen as very dangerous and should not be used. Nowadays, virtually all modern earth fault loop impedance testers have a 'no trip' option, although names may vary. When the 'no trip' earth fault loop test is selected, the tester will apply the test in a way designed not to trip the RCD and therefore the 'no trip' test function allows earth fault loop impedance testing of RCD protected circuits. Every manufacturer performs their 'no trip' test in a different, patented way. Some methods use three leads and require a neutral connection; others only use two leads, but it is generally accepted that these tests will not be as accurate as the high current tests and may be more susceptible to problems such as electrical interference or contact resistance. The tester must select the correct test method for the circuit under test and be aware of the strengths and weaknesses of the chosen test. Where doubt exists, advice should be sought from the test equipment manufacture prior to testing.

Values of earth fault loop impedance will need to be obtained at several points within the installation. These values are then recorded on the appropriate certificates and schedules for the type of inspection and testing being undertaken.

Following confirmation of correct supply polarity, the second 'live' test carried out at the origin on the installation is the external earth fault loop impedance test, often called the Z_e test. This test evaluates the quality of the earthing arrangement, which in TN systems is provided by the supplier and in the case of TT systems is provided by a combination of the installation earth electrode and the supplier's installation. On page 11 of the *IET On-Site Guide*, typical maximum values are stated for external earth fault loop impedance. These values allow us to identify where a fault may exist within the supplier's earthing arrangement.

Typical maximum external earth fault loop impedance values:

- TN-S systems 0.8Ω.
- TN-C-S systems 0.35Ω.
- TT systems 21Ω (supplier's equipment only, this value does not include the resistance of the installation earth electrode, which is typically below 200Ω).

Where doubt exists with regards to the safety of the supplier's earthing arrangement this should be brought to their attention immediately.

The external earth fault loop impedance test is performed at the origin on the installation between the incoming line conductor and the earthing conductor. To avoid readings being affected by parallel earth paths, the earthing conductor must be disconnected from the MET while the test is performed. Some METs have a removable link to allow easy disconnection. It is unsafe to disconnect the earthing conductor from a 'live' installation, so the installation must be safely isolated before disconnection can be made. The earthing conductor must only be disconnected from the MET for the shortest period of time possible to allow the external earth fault loop impedance test to be carried out and reconnection must take place immediately following completion of the test.

In addition to the tests performed at the origin of the installation, a measurement of earth fault loop impedance must be made at each distribution board. The value is commonly known as Z_s DB and, unlike the Z_e measurement above, it does not require the disconnect of parallel paths. Z_s DB is recorded on the schedule of test results. The results of the Z_s DB tests are compared with values provided by the installation's designer or to the maximum measured values applicable to the protective device protecting the distribution board.

Finally, a value of earth fault loop impedance must be obtained for each final circuit. As mentioned above this may be achieved by calculation but where this is not possible, and the levels of risk are acceptable, a 'live' measurement of earth fault loop impedance Z_s may be carried out. Using either the high current or 'no trip' test option as appropriate, measurements should be made at every point on the circuit and the highest reading recorded as the circuit Z_s on the schedule of test results.

In the case of radial circuits, the highest reading is usually expected at the furthest point of the circuit, and for ring final circuits readings will usually rise to a maximum at the centre of the ring. In the case of three-phase installations or circuits, measurements should be made between each line conductor and earth. These readings are either recorded individually or the highest reading is established and recorded.

VERIFICATION OF EARTH FAULT LOOP IMPEDANCE RESULTS

Once obtained, by whatever method, the Z_s values must be checked to ensure that they are low enough to cause the protective device to operate within the specified time if a fault occurs.

Below is an example of the calculation used to arrive at the maximum Z_s value for circuits in a TN system.

$$Zs = \frac{Uo \times Cmin}{Ia}$$

Where U_o is the nominal supply voltage to earth (i.e. 230 V), C_{min} is a new factor introduced to allow for the variation in the supply voltage (0.95 for standard installations in the UK) and I_a is the current required to operate the protective device in the specified disconnection time. I_a is established by using the graphs or tables contained in appendix 3 of the *IET Wiring Regulations* (BS 7671).

For example, if a circuit is protected by a 6A type B circuit breaker to BS EN 60898, we look at the table on page 325 of the *IET Wiring*

Regulations (BS 7671), which tells us that we need an I_a of 30A to ensure that the device will operate.

$$Zs = \frac{230 \times 0.95}{30} = 7.28\Omega$$

To save us doing this calculation every time, the *IET Wiring Regulations* (BS 7671) includes these values pre-calculated for the common protective devices in tables 41.2, 41.3 and 41.4, which start on page 57. If you look up the 6A type B circuit breaker in table 41.3 you will see that the tables shows 7.28Ω, in agreement with the example calculation above.

Although the tables in the *IET Wiring Regulations* (BS 7671) give maximum values of Z_s, unfortunately we are not allowed to use these to verify our test results. This is because there is no margin for error with these maximum values. Although the circuit would comply at the moment the reading was taken, any slight change in circuit conditions could mean that the installation may not perform correctly in the case of a fault. The main factor affecting the stability of the circuit Z_s is temperature.

Maximum ambient temperature for an installation in the UK is usually assumed to be 30°C; maximum operating temperature for conductors is generally 70°C or 90°C and testing is often assumed to be carried out at 20°C, although it could be much lower. The problem is that a measurement of Z_s, which is taken at 20°C and passes, might not pass when the circuit heats up to 30°C, 70°C or 90°C, because resistance increases with temperature. These differences in temperature must then be allowed for in the pass/fail limit specified for measured values of Z_s. This is addressed by the *IET Wiring Regulations* (BS 7671) in appendix 14 on page 452, where we get the equation below for the maximum measure value of earth fault loop impedance $Z_{s(m)}$. I have also continued our example from above to show the effect on the 6A circuit.

$$Zs(m) = 0.8 \times \frac{Uo \times Cmin}{Ia} = 0.8 \times \frac{230 \times 0.95}{30} = 5.82\Omega$$

The equation above applies a derating factor of 0.8 or 80 per cent to the maximum Z_s values from tables 41.2 to 41.4. Once corrected in this way, these values can then be used to confirm compliance with the *IET Wiring Regulations* (BS 7671) for measured values of Z_s. To make life easier, *IET Guidance Note 3* (appendix A) and *The IET On-Site Guide* (appendix B) both include tables of maximum measured Z_s ($Z_{s(m)}$), which are already derated.

To verify the Z_s values obtained for RCD-protected circuits in a TT system, we use the same maximum values as detailed earlier in Chapter 14 in the section covering the verification of earth electrode resistance test results.

CALCULATING EXPECTED Z_s VALUES FOR FINAL CIRCUITS

In your examination you may be asked to calculate the expected Z_s value for an example circuit from some basic information. Below I have included a typical example of this type of question.

Example

You are to measure the earth fault loop impedance at the furthest point of a radial circuit suppling a domestic cooker. The external earth fault loop impedance Z_e is 0.2Ω and the 20-metre cable suppling the cooker is a 6mm²/2.5mm² 70°C thermoplastic insulated and sheathed flat cable with protective conductor. Calculate, showing your workings, the expected value of earth fault loop impedance. The table below has been included for your reference.

First we start with the simple equation below.

$$Zs = Ze + R1 + R2$$

We know that the Z_e is stated to be 0.2 ohms so all we need to do is calculate $R_1 + R_2$ then we can add the values together to get the Z_s.

$$R1 + R2 = \frac{(R1 + R2) \times I}{1000}$$

Table 16.1　Calculate the expected value of earth fault loop impedance

Cable CSA (mm²)	Resistance per metre at 20°C
1.5	12.10mΩ/m
2.5	7.41mΩ/m
4.0	4.61mΩ/m
6.0	3.08mΩ/m

We need to add in the resistances per metre for the line and CPC conductor sizes specified in the question and then multiply by the length.

$$R1 + R2 = \frac{(3.08 + 7.41) \times 20}{1000} = 0.21\Omega$$

Now we have a value for the resistance of the line and CPC conductors, we can put this into the original Z_s equation.

$$Zs = Ze + R1 + R2 = 0.2 + 0.21 = 0.41\Omega$$

So we get a value of 0.41 ohms for the circuit Z_s. This is based on the assumed testing temperature of 20°C. It is worth practising this type of question as part of your revision. Try different sizes of cable and lengths. The resistance per metre data can be found in *IET Guidance Note 3* (page 128) or in the *IET On-Site Guide* (page 190).

CHAPTER 17

Testing additional protection (regulation 612.10)

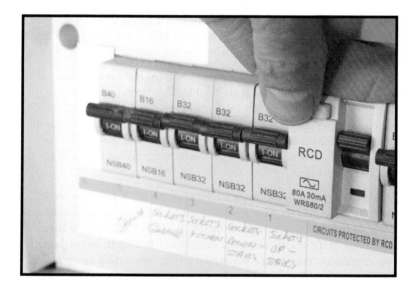

In ADS systems often additional protection is specified by the *IET Wiring Regulations* (BS 7671) or the installation's designer, to enhance protection against electric shock. Additional protection is used where there is an additional risk to the user and so therefore confirming its effectiveness is very important to the overall safety of the installation.

There are two methods of additional protection mentioned in the *IET Wiring Regulations* (BS 7671): protection by supplementary protective bonding and protection by residual current device. Where present, both must be verified by inspection and testing.

The verification of supplementary protective bonding is carried out during the continuity of protective conductors, 'dead' tests at the start of the sequence of tests, and I have described the tests in Chapter 11. The verification of RCDs, however, is a 'live' test and is normally performed as part of the functional testing at the end of the test sequence. In this section I will discuss how testing of RCDs is performed and how to check the readings obtained for compliance with the appropriate standards.

The term RCD is used to denote a residual current device, i.e. a device that operates based on the difference between the line and neutral currents (residual current). There are two common types of RCD: the RCCB or residual current circuit breaker, which operates purely on residual current, and the RCBO, which is an RCCB and circuit breaker combined, making it sensitive to both residual and overcurrents.

RCDs are also available for different applications, such as time-delayed selective RCDs (Type S), which are used to allow downstream devices to operate first and achieve selectivity between devices. Type B RCDs are sensitive to pure DC fault currents and are specified where DC fault currents may be present. It is important before testing that you know what type of RCD you have and that you have confirmed that your test equipment is suitable.

RCD testing is performed on the load side of the RCD, either at the RCD terminals directly or at a suitably accessible point on a final circuit. Loads should be disconnected during the RCD tests as they may affect the validity of the results obtained. Prior to any RCD test, an earth fault loop impedance test must be performed on the installation at the point of test to confirm that there is indeed an earth connection present. The RCD test will introduce an earth fault current into the earth path and if there is a break in this path, or high impedance, all exposed conductive parts have the potential to become 'live' during the test and pose a risk to the tester and those using the installation.

The RCD tests are covered in *IET Guidance Note 3* (page 64) and *IET On-Site Guide* (page 111), and the performance criteria are included at the beginning of appendix 3 of the *IET Wiring Regulations* (BS 7671) (page 318).

The test is performed using an RCD tester, which applies a test current to earth and then measures how long the RCD takes to disconnect the circuit. Measured values are expressed in milliseconds (mS).

The rated residual operating current of an RCD is denoted by the symbol $I_{\Delta n}$ and this value is used as the basis for the test currents. The first test applied is a test to ensure that the RCD is not too sensitive and will not operate unnecessarily. This test is known as the 'half times' test because it is performed at ½ $I_{\Delta n}$; for example, a 30mA RCD would be tested at a test current of 15mA. The result we are looking for is that the RCD does not operate. RCDs that operate may be too sensitive and further investigation should be carried out as to their serviceability.

After successful completion of the test at ½ $I_{\Delta n}$, we then perform the test again at $I_{\Delta n}$ and the RCD should now trip within 300mS (BS EN 61008 or 61009) or 200mS (BS 4293 or 7288). Time-delayed or Type S RCDs will take longer (see the *IET Wiring Regulations* (BS 7671) for the performance criteria).

Where RCDs are rated ≤ 30mA and provide additional protection against electric shock, a further test is required to also confirm compliance with the additional protection requirements of the *IET Wiring Regulations* (BS 7671). These RCDs are tested again at 5 $I_{\Delta n}$ (150mA for a 30mA RCD) and should operate within 40mS.

The results of the above tests may differ if they are performed during the positive or negative half cycles of the supply's sinusoidal waveform. To allow for this, each of the tests is performed twice, once with the RCD tester set to 0° (the positive half cycle) and again at 180° (the negative half cycle). The highest value obtained is then recorded for each test.

Table 17.1 summarises the typical tests that would be carried out on a 30mA RCD to BS EN 61008 providing additional protection.

Table 17.1   Typical tests that would be carried out on a 30mA RCD to BS EN 61008 providing additional protection

Test	Test Current	Half Cycle	Expected Results
½ $I_{\Delta n}$	15mA	0°	No trip
½ $I_{\Delta n}$	15mA	180°	No trip
$I_{\Delta n}$	30mA	0°	≤ 300mS
$I_{\Delta n}$	30mA	108°	≤ 300mS
5 $I_{\Delta n}$	150mA	0°	≤ 40mS
5 $I_{\Delta n}$	150mA	180°	≤ 40mS

Many modern multifunction/RCD testers have an automatic RCD test, which will perform all the required tests automatically and then record the values for you to review after the test. The tester simply starts the test and then resets the RCD after each test until the sequence is complete. The automatic RCD test is a real time saver, so it is worth looking to buy a tester that includes this function.

In addition to the tests performed with the RCD tester, each RCD has a built-in test button to confirm the physical operation of the device and exercise the internal parts. The test button should be pressed after the other tests have been completed to check correct operation. The *IET Wiring Regulations* (BS 7671) also require that the test button is pressed quarterly (every three months) and it is important to make the installation duty holder aware of this requirement.

CHAPTER 18

Testing prospective fault current (regulation 612.11)

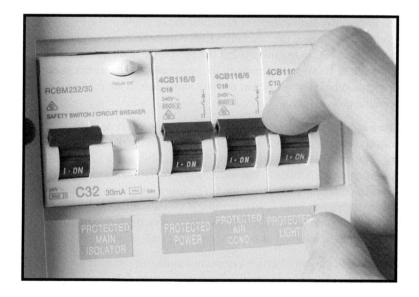

Before we can be sure that an electrical installation is safe, we must know that the protective devices are capable of handling the levels of current that could be present in the event of a fault. In some installations prospective fault currents can be tens of thousands of amps. Needless to say, many protective devices would be damaged by such high currents and may even fail catastrophically, putting the premises in danger.

We need to be able to establish a value for the maximum fault current that could be present at each protective device and verify that the device is suitably rated to cope with this current. As discussed in relation to earth fault loop impedance, it is not feasible to create a fault in the installation and then simply measure the current with an ammeter, so the test must be performed in another way. The actual method used is to measure the fault loop impedance and then convert this value into a prospective current using Ohm's law. To perform the test, we generally use the PFC range on a combined PFC and earth fault loop or multifunction tester. The values are converted and displayed by the test meter in kilo amps (kA) or in some cases amps. The prospective fault current test is a live test and therefore all live working precautions must be observed.

The prospective fault current is measured at the origin of the installation because this is where the maximum fault current will occur. As we move downstream within the installation the prospective fault current will get less. It is also usual to measure the prospective fault current at each distribution board to verify that the protective devices are capable of handling this current.

There is often much confusion about the terms used to describe prospective fault current. The terms PFC (Prospective Fault Current) or I_{pf} (Current Prospective Fault) are general terms for the maximum prospective fault current found within the installation irrespective of which conductors were measured. The more specific terms PEFC (Prospective Earth Fault Current) and PSCC (Prospective Short Circuit Current) are used to describe the individual tests; the highest of these values are then recorded as the PFC for the installation.

For a single-phase installation prospective fault current, tests are performed as below.

Table 18.1 Tests for a single-phase installation prospective fault current

Test (voltage)	Between
PEFC (230V)	Line and Earth
PSCC (230V)	Line and Neutral

For a three-phase installation prospective fault current, tests are performed as in Table 18.2 below.

Some older test equipment will not be able to carry out tests at 400V (i.e. between phases). In these cases we approximate the value by multiplying the highest single phase value obtained by $\sqrt{3}$ (1.73) or, for simplicity, the value is doubled.

Table 18.2	Tests for a three-phase installation prospective fault current
Test (voltage)	**Between**
PEFC (230V)	L1 and Earth
PEFC (230V)	L2 and Earth
PEFC (230V)	L3 and Earth
PSCC (230V)	L1 and Neutral
PSCC (230V)	L2 and Neutral
PSCC (230V)	L3 and Neutral
PSCC (400V)	L1 and L2
PSCC (400V)	L1 and L3
PSCC (400V)	L2 and L3

As with the earth fault loop impedance tests, performing PEFC measurements on RCD protected circuits will cause the RCD to operate. In the case of RCD protected circuits, the 'no trip' PFC range must be selected.

The highest of all the values obtained from the tests listed above is recorded on the test documentation as the PFC/I_{pf} for the installation.

To verify that the protective devices are suitable for the levels of PFC we must establish their rated short-circuit capacity. This information may be obtained from the manufacturer, and there is data for the most common devices included in *IET Guidance Note 3* (page 61) and the *IET On-Site Guide* (page 74). For circuit breakers to BS EN 60898 or RCBOs to BS EN 61009, two values are listed. These values are I_{cn} and I_{cs}. The I_{cn} value is the amount of current that the device can interrupt, but it may not be serviceable afterwards. In contrast, the I_{cs} value is the maximum value of current that the device can interrupt and still be serviceable afterwards. The definitions of I_{cs} and I_{cn} have been used as exam questions in the past, so try to remember I_{cn} is **n**ot serviceable and I_{cs} is **s**erviceable.

CHAPTER 19

Check of phase sequence (regulation 612.12)

In three-phase installations it is important to ensure that phase sequence is maintained from the origin through to the connection points for any three-phase current using equipment. Many items of three-phase equipment will not operate correctly if the phase sequence is wrong and this can give rise to danger. This is especially true for many three-phase motors, where reversing the phase sequence will cause the motor to rotate in the opposite direction.

Verifying correct phase sequence is a very similar process to confirming correct polarity, so first we must confirm by visual inspection that L1, L2 and L3 are correctly identified and terminated at the origin and at each point throughout the installation. Next we use a phase rotation tester to confirm the phase sequence at the origin, any three-phase distribution boards and at the connection points for any three-phase equipment.

There are two types of phase rotation tester: **rotating disc type**, which uses a small three-phase motor and gearing to rotate a disc clockwise or anti-clockwise depending on phase sequence and **indicator lamp type**, which uses electronic circuitry to check the phase sequence and then illuminate a lamp for clockwise or anti-clockwise.

CHAPTER 20

Functional testing (regulation 612.13)

The last stage before completing any inspection and testing activity is functional testing. Once the installation is reassembled and safely re-energised all equipment should be thoroughly tested for correct function. In the case of new installation this will be the first time that equipment is energised and therefore we must take extra care to check the correct operation of equipment in all modes. This is also a chance for us to confirm that we have not forgotten to reconnect items disconnected for testing purposes, often a common problem where several electricians are working together.

The correct operation of all switching, control and protective devices must be confirmed. At this stage adjustments may also need to be made to equipment with user-selectable controls. We may need to run the equipment for a period of time to confirm consistent performance or to fully check the operation of more complex equipment.

As the name suggests, we are confirming that the installation and all its component parts 'function' as expected. The best approach to functional testing is usually to start at the distribution board and then methodically energise and work through each circuit in turn. Only when we are 100 per cent satisfied that the installation performs as designed can we complete the inspection and testing activity and hand over the installation to the client.

CHAPTER 21

Verification of voltage drop (regulation 612.14)

This relatively unknown test was added to the *IET Wiring Regulations* (BS 7671) when the 17th edition was published in 2008. As a result of being a recent addition, this test sits at the end of the stated sequence of tests, which incorrectly makes it appear that it should be carried out last. Such confusion about when and how to perform this test has led to it being largely ignored by most testers and therefore many volts drop problems often go undiagnosed.

The concept of volts drop is a very simple one to understand. The distribution and final circuit live conductors have resistance and, although this resistance is only small, Ohm's law tells us that where current flows through a conductor having even a small resistance, there will be a potential difference (voltage) between the ends of that conductor.

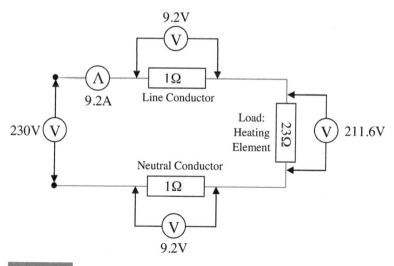

FIGURE 21.1   Diagram of voltage drop in a simple circuit

You can see the principle of volts drop illustrated in Figure 21.1. To keep things simple, in this example I have chosen as the load a 2.3kW heating element that has a resistance of 23Ω and line and neutral conductors each a having a resistance of 1Ω.

You can see that in the example circuit in Figure 21.1 that the volts drop is 18.4 Volts or 8 per cent – well above the 5 per cent limit specified for this type of circuit in the *IET Wiring Regulations* (BS 7671). BS 7671 sets out volts drop limits in appendix 4, table 4Ab on page 338. Essentially, between the origin of the installation and any load point, the percentage voltage drop should not exceed 3 per cent for lighting circuits and 5 per cent for other uses. The voltage drop is calculated as a percentage of the nominal supply voltage.

Regulation 612.14 suggests two options for verification of voltage drop: measurement or circuit impedance, or calculation. Calculation is naturally the easier option and for new installations where this work have already been done by the installation's designer, no additional inspection and testing is required to confirm volts drop beyond confirming that the installation work has been carried out in accordance with the original design. In existing installations, however, it is likely that no design information is available and there may have also been additions and alterations carried out during the life of the installation that will make the original volts drop calculations obsolete. In these cases it will be necessary for compliance with the volts drop requirements to be verified by testing.

The verification of volts drop test is performed in much the same way as the test to confirm protective conductor continuity by the $R_1 + R_2$ method, however in this case we are concerned with the resistance of the live conductors. The measurement is performed as a dead test with the installation (or, where appropriate, individual circuit) safely isolated from its supply. With all loads disconnected, a temporary link is placed between the live conductors under test at the distribution board and a measurement is made between these conductors with a low resistance ohmmeter at all points on the circuit. The highest reading obtained is taken to be the live conductor resistance, often called $R_1 + R_n$ in the case of a single-phase circuit.

$$\text{Volts drop} = (R_1 + R_n) \times \text{mf} \times I_b$$

In the equation above you can see that we must multiply the measured value obtained by a temperature multiplication factor to correct the value for the cable's maximum operating temperature. The multiplication factors from *IET On-Site Guide* table I2 on page 191 can be used if measurements were taken at 20°C. Next the value is multiplied by the design current I_b. If this is not known, the nominal rating of the circuit protective device I_n can be used as a worst case alternative.

The above equation will produce a result in volts which must be that calculated as a percentage of the nominal supply voltage before we can check compliance against the percentage voltage drop limits from BS 7671. Where any doubt exists in relation to voltage drop, further investigation should be carried out to establish the true characteristics of the circuit and the load, and the inspector must satisfy themselves that the effects of voltage drop will not impede the safe operation of the installation or any installed equipment.

CHAPTER 22

Inspection and testing certification

Recording our inspection and testing activities is vitally important for many reasons: records provide evidence that the activity took place and detail the results; records are used by future electricians to identify alterations and additions and to track deterioration; records also provide valuable feedback to the client about the safety of the installation for continued service. Although it feels like an extra job filling in all those forms, *all* electrical work must be inspected and tested and the proper records kept.

The *IET Wiring Regulations* (BS 7671) contains example forms in appendix 6, which starts on page 413, and both *IET Guidance Note 3* and the *IET On-Site Guide* have completed examples of these forms too. While the BS 7671 forms are the basis for most formal training courses, assessments and examinations, many trade organisations or companies have their own versions based on the BS 7671 forms, but altered to better meet the needs of the particular organisation. While the use of other forms is allowed, as long as they contain no less information than the BS 7671 forms, I will be concentrating on the BS 7671 forms in this book.

The forms essentially break down into two sets, those used for initial verification and those used for periodic inspection and testing.

INITIAL VERIFICATION FORMS

- Electrical Installation Certificate (EIC) and,
- Schedule(s) of Inspections and,
- Schedule(s) of Test Results

OR

- Minor Electrical Installation Works Certificate.

Typically, for new installations, the addition of a new circuit or the replacement of a distribution board, the person performing the initial verification will complete an electrical installation certificate, at least one schedule of inspections and at least one schedule of test results. More schedules can be added for larger or more complex installations.

Where the installation work does not extend to the addition of a new circuit, it may be deemed as 'minor works', for example the addition of a new socket outlet to an existing circuit. In this case we can use a single minor electrical installation works certificate to record the essential safety tests performed on this installation. If minor work is carried out on more than one circuit, then a minor electrical installation works certificate is required for each circuit. Alternately, the person performing the inspection and testing may opt for an electrical installation certificate, schedule of inspection and a schedule of test results instead.

PERIODIC INSPECTION AND TESTING FORMS

- Electrical Installation Condition Report (EICR) and,
- Schedule(s) of Inspections and,
- Schedule(s) of Test Results.

When performing periodic inspection and testing, we are required to complete an Electrical Installation Condition Report (EICR), previously known as a Periodic Inspection Report (PIR), along with at least one schedule of inspection and at least one schedule of test results. Again, where required we can use additional schedules for larger or more complex installations.

When completing inspection and testing forms, we must remember that these are effectively 'legal documents' that we will sign, and they could be used later in court if things go wrong. Always take the time to complete the forms in as much detail as possible and record all information accurately. Things that may seem obvious to you at the time may not be clear many years down the line or when read by others. A minor electrical installation works certificate that states 'replaced socket in bedroom' is a typical example. Which socket? And for another thing, which bedroom? The quality of our paperwork reflects heavily on our professionalism as electricians. Often the first tell-tale sign of a bad installation is bad paperwork.

Once completed, copies of the forms are given to the person ordering the works (client) and any other relevant parties, such as insurance companies or licensing authorities.

Correctly completing the forms associated with inspection and testing activities is an essential part of being competent. You should practise filling out sample forms until you become confident and understand every box. During the practical assessments for formal inspection and testing training courses, you will be asked to correctly fill in the appropriate forms. It has been known for form-based questions to also pop up in written/multiple-choice exams too. I am always surprised when candidates fail assessments due to not being able to complete the forms correctly or by running out of time because the paperwork took too long. There is no excuse for this, as the forms are all freely available and practising the paperwork is always covered during practical training sessions.

Part 3

CHAPTER 23

Good examination practice and revision

With proper preparation everyone has the ability to pass any inspection and testing assessment. Whether it is a written exam, on-line short answer, on-line multiple-choice or practical assessment, the key is always preparation. The old saying 'fail to prepare, prepare to fail' could have been written for inspection and testing assessments.

Proper preparation starts way back when we select our course in the first place. By selecting achievable goals we set the ground work for good assessment performance. When starting any training course one of the first things you must do is understand how the course will be assessed. You can then focus your learning on the end goal of passing the assessments. I see many candidates who assume that just by sitting in a classroom for several days, magically they will have absorbed the right information to pass the assessments. These passive learners rarely do well; only by actively engaging with the training process can you increase the chance of passing. The best learners are actively involved in perusing the knowledge they need to pass the assessments, inside and outside of the classroom. They take responsibility for their own learning and will only take the assessments when they are ready and have a good chance of passing. In this chapter I have attempted to provide guidance about how to prepare for your assessments to give you the best chance of passing and achieving your goal.

EXAM PREPARATION

Modern inspection and testing exams can take three different forms: on-line multiple-choice, on-line short answer, and written. In the past the inspection and testing examination was a formal written exam, where the candidate was given a question paper and had to write the answers out long hand in an answer booklet made up of sheets of lined paper. Many candidates struggled with this format and failed because they were not able to get their knowledge down on paper under exam conditions. The other main drawback of written exams was the long turnaround time for marking each exam, getting the results and then if necessary taking a re-sit, all of which could easily take several months. Now most qualifications have moved to a more flexible approach by utilising on-line assessments. These qualifications use either a multiple-choice exam alone or the combination of a multiple-choice exam with a short-answer format for the higher level awards.

On-line assessments have the advantage of a more structured format and turnaround times are now down to a matter of days. Although on-line assessments do add a degree of difficulty for those with limited IT skills, generally they are seen as a much easier option when compared with the old written exams. If you are not comfortable using a computer get some practice at home using the computer of a friend or family member. Practise using the mouse and keyboard and become comfortable with navigating your way around simple menus. A little practice at home will go a long way to alleviating the fear of using a computer in an exam scenario.

Remember, the exam system has been designed to be easy to use and an invigilator will always be on hand to help you if you get stuck. Before your exam you will also get the chance to undergo a brief tutorial on how to use the exam system and to practise the various tasks involved. Some awarding organisations also have sample on-line exams that you can try as part of your revision, so always find out what is available to help with your preparation.

Any candidates with special requirements or learning difficulties should make these known to the examination centre as early as possible, ideally before enrolling on the course. This will give the examination centre time to make any necessary arrangements and ensure that your needs are fully met on the day. If you are not sure what help may be available, speak to your examination centre.

In your exam you are able to use a non-programmable calculator, so ensure you take one along and that you have practised using it to perform the type of calculations that you may encounter during your exam. You are also allowed a blank sheet of paper for calculations and making notes during the exam.

ON-LINE MULTIPLE-CHOICE EXAMS

Once the exam begins, address each question one at a time, read them carefully and select the appropriate answer. Flag any questions you are

unsure of. You are able to change your mind at any point during the
exam and your answers are not set until the time ends. Displayed at
the top of your examination screen will be an indication of the time
remaining and your progress through the exam. Keep an eye on the time
so that you can speed up or slow down as necessary, but try to leave a
little time at the end to look back over any flagged questions. There are
no marks for finishing early, so try to use all the time available and pace
the exam so that you do not have to rush at the end.

WRITTEN OR ON-LINE SHORT-ANSWER EXAMS

Again, the key here is to read the question carefully, including all the b,
c, d, parts. Reading the question in full will allow you to structure your
answer properly and often you will get some clues from the way the
question is divided up. Take great care to use correct terminology – you
may need to practise this as part of your revision. Simply confusing
terms like 'live' and 'line' can totally change the meaning of your answer
and result in low marks.

The awarding organisation City & Guilds has a 'Chief Examiner's
Reports' section on their website that gives guidance about how
candidates have performed on previous exams and the typical mistakes
that are made. This type of information is very useful in preparing
yourself for assessment so well worth a look. Time management is
key to this type of exam and always keep an eye on the clock. As with
everything, 'practice makes perfect' and the more you can practise
writing or typing your answers during revision, the easier the exam will
seem and the more time you will feel you have.

PRACTICAL ASSESSMENTS

As I mentioned earlier, most people fail the practical assessment because
they run out of time, not because time is tight for the assessment but
because they have not sufficiently prepared and therefore everything
takes twice as long. If possible you should buy or borrow an electrical
installation tester and practise using it as much as possible before
the assessment. Most assessment centres will allow you to use your
own tester during the assessment, but this is often an advantage most
candidates ignore. On many occasions I have heard candidates make
the excuses that they usually use another brand of tester or don't know

how to use the tester because they only bought it yesterday. These are amateur mistakes that could be avoided. The same is also true for completing the inspection and testing paperwork, where there is no excuse for not being fully prepared.

The contents of all practical assessments follow the same format. You will be asked to perform inspection and testing activities on a simulated electrical installation and complete the appropriate paperwork. Your preparation should focus on: safety and safe isolation procedures; being able to perform visual inspection based on the schedule for inspections; being able to perform the complete range of 'dead' and 'live' tests and complete the associated forms.

Where possible, and when safe to do so, you should practise the above before the assessment. Often candidates use their own home or build a test rig to allow them to practise. Some candidates even have the opportunity to shadow a colleague or friend while they perform inspection and testing in the real world. Anything that builds up your confidence, fluidity and speed during testing will greatly improve your chances of passing with ease.

AFTER THE EXAM OR ASSESSMENT

Your results will usually be available straight away, but for written/short-answer exams there will be some delay. The exam centre will be able to give you a rough idea of when to expect the results. You will receive your results along with any feedback, which can be useful in highlighting strengths and weaknesses. If you have failed the exam/assessment, look at your performance and try to identify in which areas you were weak so that you are able to concentrate on those areas when revising for your re-test.

It is common for many candidates to fail the exam on their first try, so do not feel too despondent if you don't pass straight away. If you do fail, do not rush to book a re-test straight away. Allow a suitable time period for further training and revision to ensure you have a better chance next time. Only re-sit when you are ready to do so and have a good chance of passing.

CHAPTER 24

Sample examination questions

I have included below 50 sample examination questions, which you can use as part of your revision. You must be aware, however, that memorising these sample questions will not help you pass the exam; they are intended to give you exam practice and show you exam content and how the topics covered earlier in this book may be addressed. Your revision should concentrate on the subject matter and the guidance material, which will then ensure that you are able to answer the full range of questions during your exam. At the end of the chapter I have also included answers to the sample questions for you to refer to.

SAMPLE REVISION TEST

1. What is the most likely human sense that would reveal a loose bush in a steel conduit installation?
   (a) Touch
   (b) Hearing
   (c) Taste
   (d) Sight

2. A person who carries out inspection and testing is legally referred to as a(n):
   (a) Approved person
   (b) Responsible person
   (c) Competent person
   (d) Authorised person

3. The resistance of a conductor is **not** affected by the:
   (a) Conductor CSA
   (b) Insulation material
   (c) Ambient temperature
   (d) Conductor length

4. When insulation resistance is measured at a distribution board with additional final circuits connected, the overall insulation resistance will:
   (a) Be lower
   (b) Be higher
   (c) Be unchanged
   (d) Be halved

5. A check for polarity is carried out at the origin of the installation. This is to confirm:
   (a) Operation of the main isolator
   (b) The type of earthing system
   (c) Correct connection of the incoming supply
   (d) The rating of the supply fuse

6. When carrying out a test for prospective short circuit current, between lines, in a three-phase installation, the prospective fault current tester must be rated for a nominal voltage of:
   (a) 400VAC
   (b) 230VAC
   (c) 500VAC
   (d) 100VAC

7. The units displayed on an earth continuity and insulation resistance tester are:
   (a) $M\Omega$ and kVA
   (b) Ω and $M\Omega$
   (c) $m\Omega$ and A
   (d) Ω and kA

8. Which of the following measures for protection against electric shock may be found in a location containing a bath or shower?
   (a) Earth-free local equipotential bonding
   (b) Placing out of reach
   (c) SELV
   (d) Non-conducting location

9. The expected earth fault loop impedance of a circuit may be calculated using the formula:
   (a) $Z_e = Z_s + (R_1 - R_2)$
   (b) $Z_s = Z_e - (R_1 + R_2)$
   (c) $Z_e = Z_s - (R_1 + R_2)$
   (d) $Z_s = Z_e + (R_1 + R_2)$

10. For safety reasons what test must always precede an RCD test?
    (a) A prospective fault current test
    (b) A phase rotation test
    (c) A test to verify voltage drop requirements
    (d) An earth fault loop impedance test

11. Which of the following insulation resistance tests would be suitable for a 230VAC circuit containing surge-protected equipment that cannot be removed?
    (a) Test voltage 250V minimum resistance 0.25MΩ
    (b) Test voltage 250V minimum resistance 1.0MΩ
    (c) Test voltage 500V minimum resistance 0.5MΩ
    (d) Test voltage 500V minimum resistance 1.0MΩ

12. Where RCDs are to be connected in series, to ensure correct operation in the event of a fault there must be:
    (a) Suitability
    (b) Selectivity
    (c) Switchability
    (d) Shareability

13. The most important precaution to be taken before disconnecting the earthing conductor for test purposes is:
    (a) Seek permission from the DNO
    (b) Check for parallel paths
    (c) Remove the earth electrode
    (d) Isolate the supply and lock off

14. The correction factor to be applied when making comparison between measured and maximum tabulated values of earth fault loop impedance is:
    (a) 80 per cent
    (b) 70 per cent
    (c) 90 per cent
    (d) 75 per cent

15. When a test of insulation resistance is carried out, any two-way or intermediate switching must be:
    (a) Isolated from the circuit
    (b) Clearly labelled
    (c) Tested in all switching positions
    (d) Double insulated

16. The type of system that uses the general mass of earth as the earth return path to the transformer is:
    (a) TT
    (b) TN-S
    (c) TN-C-S
    (d) IT

17. The earth lead of the earth fault loop impedance tester used to carry out a test of external earth fault loop impedance must be attached to:
    (a) Main earthing terminal
    (b) Main protected bonding conductors
    (c) Disconnected earthing conductor
    (d) Circuit protective conductors

18. The measured value of external earth loop impedance for an installation is 0.35 ohms. If the measured $R_1 + R_2$ value for one of the circuits is 0.7 ohms, the expected Z_s for the circuit is:
    (a) 2.10 ohms
    (b) 0.95 ohms
    (c) 1.05 ohms
    (d) 0.22 ohms

19. According to *HSE Guidance Note GS38*, test probes for use on low voltage circuits should have:
    (a) No more than 4mm exposed metal at the tip
    (b) No more than 3mm exposed metal at the tip
    (c) No more than 2mm exposed metal at the tip
    (d) No more than 1mm exposed metal at the tip

20. The relationship between the maximum prospective fault current and the breaking capacity of any protective device is that the breaking capacity:
    (a) Must exceed the maximum PFC
    (b) Must not exceed the maximum PFC
    (c) Must only exceed the maximum PFC for radial circuits
    (d) Must not exceed the maximum PFC for radial circuits

21. One method of evaluating voltage drop listed in BS 7671 is by:
    (a) Measuring circuit impedance
    (b) Using a volt meter
    (c) Measuring the supply voltage
    (d) Measuring earth fault loop impedance

22. The maximum BS 7671 tabulated value of earth fault loop impedance
    for a circuit is 1.44Ω. If the resistance of the line and CPC of the final
    circuit is 0.95Ω, the maximum acceptable value of external earth fault
    loop impedance is:
    (a) 0.82Ω
    (b) 0.49Ω
    (c) 0.35Ω
    (d) 0.20Ω

23. The test voltage stated in BS 7671 for a test of insulation resistance on
    a circuit operating at 230VAC is:
    (a) 250VDC
    (b) 250VAC
    (c) 500VDC
    (d) 500VAC

24. The results of a test for ring final circuit continuity are all satisfactory
    apart from one socket outlet that reads open circuit when tested during
    step two. This may indicate:
    (a) A bad neutral connection at the socket outlet
    (b) A bad line connection at the socket outlet
    (c) A bad CPC connection at the socket outlet
    (d) Reversed polarity between neutral and line at the socket outlet

25. If the length of a copper conductor is double, its resistance will:
    (a) Quadruple
    (b) Double
    (c) Stay the same
    (d) Halve

26. An RCD is installed to provide additional protection. What is the
    maximum acceptable operating time when the device is tested at the
    highest required test current?
    (a) 30ms
    (b) 200ms
    (c) 40ms
    (d) 300ms

27. Any item excluded from the inspection process by the agreement with both the inspector and the client should be marked:
    (a) FI
    (b) N/A
    (c) ✓
    (d) LIM

28. Which of the following do **not** require a warning notice stating 'safety electrical connection do not remove'?
    (a) A main earthing terminal separate from the main switchgear
    (b) The connection of an earthing conductor to an earth electrode
    (c) The connection of a main protective bonding conductor to an extraneous conductive part
    (d) The connection of a CPC to an exposed conductive part

29. When certifying a new installation that extends to several circuits, an electrical installation certificate may be:
    (a) Replaced by a minor electrical installation works certificate
    (b) Reproduced from BS 7671 only if it is exactly as published
    (c) Reproduced in any form as long as it contains no less information that the BS 7671 certificate
    (d) Replaced by an Electrical Installation Condition Report

30. Hearing is the most likely human sense to indicate:
    (a) A worn bearing on a motor
    (b) An overloaded bus bar
    (c) A missing warning label
    (d) Sharp edges on metal trunking

31. One part of the earth fault path for a TN-C-S system is:
    (a) PEN conductor
    (b) General mass of earth
    (c) Neutral conductor
    (d) The sheath of the supply cable

32. Guidance about the safe design of electrical test equipment for use by electricians is found in:
    (a) BS 7671
    (b) The Electricity at Work Regulations 1989
    (c) HSE Guidance Note GS38
    (d) IET Guidance Note 1

33. An Initial Verification should be carried out:
    (a) Every ten years for domestic properties
    (b) During erection, on completion and before it is put into service
    (c) At an interval specified by the designer
    (d) Where specified by an insurance company

34. SELV equipment located in zone 1 of a location containing a bath or a shower must have a maximum AC voltage of:
    (a) 50V
    (b) 25V
    (c) 12V
    (d) 6V

35. The earthing conductor connects:
    (a) An exposed conductive part to an extraneous conductive part
    (b) The main earthing terminal to the means of earthing
    (c) An exposed conductive part to the main earthing terminal
    (d) Structural steelwork to the main earthing terminal

36. Before performing initial verification on a new electrical installation, the inspector must establish:
    (a) The type of earthing system
    (b) The metering company
    (c) The age of the energy meter
    (d) The type of supply cable

37. When performing a polarity test on a final circuit in a three-phase installation, what should be the voltage between line and earth?
    (a) 0VAC
    (b) 120VAC
    (c) 230VAC
    (d) 400VAC

38. A three-signature electrical installation certificate must **not** be signed by the:
    (a) Designer
    (b) Inspector
    (c) Customer
    (d) Constructor

39. What additional form(s) are required to accompany an electrical installation certificate?
    (a) Schedule of inspections only
    (b) Schedule of test results only
    (c) Both a schedule of inspections and a schedule of test results
    (d) A minor electrical installation works certificate

40. Under the requirements of the *Electricity at Work Regulations*, persons carrying out an inspection and test on an installation must ensure the safety of:
    (a) Themselves
    (b) Others and the customer
    (c) Both themselves and others
    (d) Your staff

41. The electrical installation is isolated while performing an external earth fault loop impedance test because:
    (a) The installation will have no earth during the test
    (b) The exposed conductive parts may become live
    (c) Any connected equipment will affect the test results
    (d) The test may damage the main earth terminal

42. What is the typical U_o in a single-phase reduced low voltage system?
    (a) 230VAC
    (b) 55VAC
    (c) 110VAC
    (d) 400VAC

43. Before conducting an inspection and test of a new cooker circuit, one of the most important items of information an inspector should have available is:
    (a) The type and rating of the protective device
    (b) The manufacturer of the cooker
    (c) The date on which the cooker was originally installed
    (d) Warranty details for the cooker

44. The unit of measurement for insulation resistance is:
    (a) $m\Omega$
    (b) Ω
    (c) $k\Omega$
    (d) $M\Omega$

45. The user of the installation should be advised before a test of insulation resistance is carried out because of the danger from:
    (a) The test voltage
    (b) Damage to equipment
    (c) Loss of supply
    (d) Disruption to services

46. Which of the following documents relates most closely to inspection and testing of an electrical installation?
    (a) BS 7671
    (b) IET Guidance Note 3
    (c) The Electricity at Work Regulations
    (d) HSE Guidance Note GS38

47. An alternative method used to establish the value of prospective short circuit current for a three-phase installation when direct measurement is not possible is by:
    (a) Doubling the single-phase value
    (b) Halving the single-phase value
    (c) Using the 80 per cent rule
    (d) Adding together the PEFC and the PSCC

48. Step three of a test of ring final circuit continuity requires us to:
    (a) Measure the end-to-end resistance of the RFC conductors
    (b) Cross connect the line and neutral conductors
    (c) Cross connect the line and CPC
    (d) Cross connect the neutral and CPC conductors

49. If a test instrument is connected across line and CPC and its scale is ohms, what is it likely to be connected to measure?
    (a) Continuity of protective conductors
    (b) Insulation resistance
    (c) Live polarity
    (d) Earth electrode resistance

50. The accessible top surface of an enclosure must be rated:
    (a) IPX4
    (b) IP4X
    (c) IP2X
    (d) IPX2

ANSWERS

Table 24.1   Answers to revision questions

Question	Answer	Question	Answer
1	A	26	C
2	C	27	D
3	B	28	D
4	A	29	C
5	C	30	A
6	A	31	A
7	B	32	C
8	C	33	B
9	D	34	B
10	D	35	B
11	B	36	A
12	B	37	C
13	D	38	C
14	A	39	C
15	C	40	C
16	A	41	A
17	C	42	B
18	C	43	A
19	A	44	D
20	A	45	A
21	A	46	B
22	B	47	A
23	C	48	C
24	A	49	A
25	B	50	B

Index

Note: Page numbers in **bold** type refer to **figures**
Page numbers in *italic* type refer to *tables*

A-Levels 20
accompaniment 29
accurate equipment 66–8, **67**
additional protection testing 121–4, *123*
all systems 47
ammeter 125
apprenticeship 18
Approved Document P (Part P) 14, 20
area, cross-sectional (CSA) 86–7
assessment: practical 57, 143–6; risk 28–30, 32, 38, 44, 91
automatic disconnection of supply (ADS) 71, 76, 107–12, 121
available qualifications 22–4, *22*

basic protection 89
Best Practice Guide (2) 32–5, 51
bonding clamp 109
BS EN (61010) 36
BS EN (61243-3) 35–6
building control body: certification 14; local 14–15
Building Regulations 4
building regulations 14–15, 20
Building Regulations (UK, 2010) 42

cables: connected in parallel 95–7; length 95; LV 112
calibration 67
CAT (Category) ratings 37
CAT IV 100–1
certification 14; building control body 14; EIC 137–8, 152–4; inspection and testing 137–9; Minor Electrical Installation Works 137; self 14–15; third-party 14–15

change of ownership 47
change of use 48
Chief Examiner's Reports (City & Guilds) 145
circuit conductors 81
circuit protective conductors (CPC) 71, 75–85, 119, 151–2, 155; continuity of 76–80, **79**
circuit protective device 135
circuits 34, 39, 80–1, 95, 108, 114; change in conditions 117; earth fault loop impedance 77–9; existing 138; expected values for final 118–19; faulty 97; final 148; isolated 38; radial 116; single phase 134; testing 95; three phase 69
City & Guilds (2391) 19–21
City & Guilds (2392) 20–1
City & Guilds (2394) 21
City & Guilds (2395) 21
City & Guilds (UK) 19–23; Chief Examiner's Reports 145
clamp type tester 104
classification codes 52, **52**
clothing 28
Code (C1) 51–2, 58–9
Code (C2) 51–2, 58
Code (C3) 51–2, 59
codes, IP 59–61, *61*
Cold War (1945–92) 20
college-based training course 18
combined neutral and earth conductor (CNE) 109
common sense 32, 55
common testing practice 104
competency 13–14, **13**
competent person 12–15; schemes 14–15
complex installations 23, 69, 138

conductors 33, 71–88, 99, 117, 133;
circuit 81; circuit protective bonding
71, 75; circuit protective (CPC) 71,
75–85, 119, 151–2, 155; combined
neutral and earth (CNE) 109;
continuity testing 71–88; copper
151; earthing 71–3, **72**, 115, 149,
153; individual 76; length 87–8;
line 79–80, 83–4, 94, 100, 115–16,
134; live 32, 75, 82, 92–4, **93**,
133, 134; main protective bonding
71–5, 87; neutral 83–4, 94, 108,
134; protective 72, 86–7, 99, 108,
118, 121; protective earth and
neutral (PEN) 109–10; reasonable
33; resistance 86–8, 147; ring final
circuit 81–3, **81**, **83**, **85**, 86, 99;
supplementary protective bonding
71, 75–6; testing continuity of 71–6;
uninsulated 32–3; unreasonable 33
control: of risks 28; of test areas 28
control measures 30, 114
controls, user-selectable 131
copper conductor 151
correct phase sequence 129
cowboy electricians 4
cross connection: of line and CPC
84–6; of line and neutral 83–4
cross-sectional area (CSA) 86–7
current 125–7; short circuit 148
current prospective fault (I_{pf}) 126
customers 30, 46

damage 48
DC voltage 92; test 92, *92*
dead testing 8, 63–4, 71, 90, 99–101,
134, 146
defects 50
disconnection: automatic 71, 76,
107–12, 121; time 113
DIY 4, 46, 99, 108
domestic installations 20
domestic sector 15
dual coil clamp test 104
Duty Holder 13, 27, 33
dwellings 14–15

E1 test 103–4
E2 test 104–5
E3 test 105

EAL (UK) 19–23
earth electrode resistance testing 65–6,
103–6, 118
earth electrode (TT systems) 35, 72,
103–6, 107, 110–12; diagram **111**;
identification **111**; single 104
earth fault loop impedance 148–51;
circuit 77–9; external 150
earth fault loop impedance test 105,
107, 113–19, 127; expected value
118; external 77, 114, 154; live
114; no trip 114; results verification
116–18
earth fault path 113, **113**
earth return path 111
earth terminals 99
earthed metallic wiring systems testing
80
earthing: arrangement 115; main
terminal (MET) 72, 77–9, 90–1,
100, 115; means (TN systems) 72
earthing conductor 71–3, 115, 149,
153; continuity **72**
earthing systems 105, 107–8, 148;
definitions *108*; letter codes 107,
107
education 12
ELECSA (UK) 15
electric shock 121–3, 148
electrical hazards 12
electrical industry 3, 11, 18–20
Electrical Installation Certificate (EIC)
137–8, 152–4
Electrical Installation Condition
Reports (EICRs) 23, 47–53, 59, 138,
152
electrical installations *see* installations
electrical resistance 90
electrical safety 17
Electrical Safety Council 32
Electrical Safety First, *Best Practice
Guide (2)* 32, 42, 51, 68
electrical systems 47
electrical theory 17–18
electricians 5, 8, 18–20, 31–2, 76,
114; cowboy 4; maintenance 4;
professional 15; unqualified 99
electricity 12, 63
Electricity at Work Regulations (1989)
4, 12–13, 31–3, 42, 51, 64, 154

electricity boards, local 3
Electricity Safety Quality and
    Continuity Regulations (UK, 2002)
    42
electricity supply 34; network 105
electrodes 106
employees 13
employers 13–14, 18–19
end-to-end measurements 81–3
England 14
examination: practice 143–6, 147;
    preparation 143–4; system 144;
    written 143–5
existing circuit 138
existing installations 8–9, 42–4, 46–8,
    57, 134
extent and limitations declaration
    49–50
external earth fault loop impedance
    test 77, 114, 154
external earth loop impedance 150

fault loop impedance 126
fault protection 89
faulty circuit 97
final circuits 148; continuity 151;
    expected values 118–19; live
    conductors 133
fixed installation 37
formal inspection and testing
    qualifications 19
formal training courses 137
functional testing 8, 131

generators 103
Great Britain 13

hazards 30, 32, 63; electrical 12
health and safety 4, 14, 18, 27–30,
    41, 44
Health and Safety at Work Act (UK,
    1974) 42
*Health and Safety Executive Guidance
    Document HSR (25)* 32, 42
*Health and Safety Executive Guidance
    Note (GS38)* 27, 36, **36**, 42, 66, 150
high current 114; test 114
houses in multiple occupation (HMOs)
    48
human senses 55–6, 152

IET Guidance Note (1) 60
IET Guidance Note (3) 17–19, 56–7,
    66–8, 76–9, 103–4, 118–19
IET On-Site Guide 19, 25, 42, 118–19,
    127, 137
IET Wiring Regulations 3–5, 17–21,
    35–7, 48–52, 56–62, 121–4,
    133–4
indicator lamp types (phase rotation
    tester) 130
individual conductors 76
information: adequate 28–9; non-
    statutory 41–3; required 41–4;
    statutory 41–3; suitable/sufficient
    44; testing 43–4
initial inspection 57
initial verification 8–9, 18–23,
    42–3, 45–53, 56–9, 63–5, 89–90;
    examination 57; forms 137; testing
    requirements 64
inspecting electrical installations
    55–62
inspection: initial 57; modern 143;
    periodic 8–9, 18–23, 43–4, 45–53,
    63–4, 69–70, 138–9; scheduled
    137–8; visual 55, 99–101, 146
inspectors 9, 27–30, 49–53, 56–62, 69,
    152–4
installations 3–4, 27–9, 34–7, 47–50,
    67–8, 154–5; of ADS 71, 76, 121;
    complex 23, 69, 138; domestic 20;
    inspecting 55–62; live 103, 115;
    low-voltage 32; new 8, 42–4, 45–6,
    56–8, 92, 134, 152–3; older 92;
    original 46; process 8; safety and
    correct functioning 7; single minor
    138; special 62, **62**; tester 145;
    testing 63–70
insulation 89, 95
insulation resistance 94–5, 95–7,
    148, 151–5; affecting factors 95–7;
    minimum values 92, *92*; parallel 96;
    testing 89–97, 148–9
insurance 48
IP (International Protection) Codes
    59–61, *61*; meanings 60, *61*
isolated circuit 38
isolation: activity 38; planned 38, *see
    also* safe isolation
IT: skills 144; systems 35

knowledge 143; technical 13–14, 17–18

LCL (UK) 19–23
legal requirements 41–2
licensing 48
lighting, adequate 29–30
limitations 49; declaration 49–50; operational 50
line conductor 79–80, 83–4, 94, 100, 115–16, 134
line terminals 99
live conductor 32, 75, 82, 92–4, 134; final circuit 133; measurement of insulation resistance **93**; resistance 134
live earth fault loop impedance test 114
live installation 103, 115
live polarity test 101
live testing 8, 28–30, 63–4, 99, 105, 122, 146
live working precautions 39
loading, increased electrical 48
local electricity boards 3
low resistance measurement 74–5, 78
low resistance ohm meter 71–7, 81, 84
low-voltage installations 32
LV cables 112

main earthing terminal (MET) 72, 77–9, 90–1, 100, 115
main protective bonding conductor 71–5, 87; continuity **74**
maintenance 47; activity 46–7; electricians 4; requirements 47
material 86
maximum voltage 37
measurement: of insulation resistance between live conductors **93**; of insulation resistance between live conductors and earth **93**; resistance 72, 74–5, 75–7, 78, 81, 84
mega ohms 92
metal conduit system 57
metallic wiring system 80
Minor Electrical Installation Works Certificate 137
multiple-choice assessment, on-line 143–5

NAPIT (UK) 15
neutral conductors 83–4, 94, 108, 134
neutral terminals 99
new installations 8, 42–4, 45–6, 56–8, 92, 134, 152–3
NICEIC (National Inspection Council for Electrical Installation Contracting, UK) 4, 15
no trip earth fault loop test 114
nominal supply voltage 134–5
nominal voltage 148
non-compliances 50
non-statutory information 41–3

observations 50–3
OFQUAL (Office of Qualifications and Examinations Regulation, UK) 19–22
Ohm's law 126, 133
older installations 92
on-line multiple-choice assessment 143–5
on-line short answer assessment 143–5
operational limitations 50
original installation 46
overvoltage category 36–7

parallel current paths 71–3, 76–9
parallel insulation resistances 96
parallel paths 105
parallel resistance 87, 96
periodic inspection 8–9, 18–23, 43–4, 45–53, 63–4, 69–70, 138–9
Periodic Inspection Report (PIR) 138
periodic insulation resistance testing 89
periodic testing 89; requirements 64–5
phase rotation tester 130; indicator lamp type 130; rotating disc type 130
phase sequence 129–30; correct 129
planned isolation 38
polarity 101; faults 99–101
polarity test 99–101, 153; live 101; for single phase installation 100, *100*; for three phase installation 101, *101*
PPE test equipment 100
practical assessment 57, 143–6
practical electrical work 18
practical skills 12

practical training sessions 139
precautions 91, 94
preconditions 91, 94
professional electricians 15
prospective earth fault current (PEFC) 126–7
prospective fault current (PFC) 126–7, 150; testing 125–7
prospective short circuit current (PSCC) 126
protection: additional 121–4, *123*; basic 89; by ADS 71, 76, 107–12, 121
protective conductor continuity 134; circuit 76–80, **79**
protective conductors 72, 86–7, 99, 108, 118, 121; circuit (CPC) 71, 75–85, 119, 151–2, 155
protective devices 70, 125–7, 131, 150; circuit 135
protective earth and neutral conductor (PEN) 109–10
protective equipment 35
protective multiple earth (PME) 110–12

$R_1 + R_2$ method 76–80
R_2 (wander lead method) 76–8
radial circuits 116
reasonable conductors 33
regulations: building 4, 14–15, 20, 42; wiring 9, 18
requirements 11; legal 41–2; maintenance 47; safety 3, 12
residual current circuit breaker (RCCB) 122; with overcurrent protection (RCBO) 122, 127
residual current devices (RCDs) 69, 105–6, 112, 114, 122–4, 127, 149–51
resistance: conductor 86–8, 147; live conductor 134, *see also* insulation resistance
resistance measurements 72, 75–7, 81, 84; low 74–5, 78
revision 143–6
ring final circuit 81; continuity 155
ring final circuit conductors 81–3, **81**, **83**, **85**, 86, 99; continuity 80–6
risk assessments 28–30, 32, 38, 44, 91

risks, control of 28
rotating disc type (phase rotation tester) 130

SADCOWS (Safety, Ageing, Damage, Corrosion, Overloading, Wear and Tear, Suitability) 57, **58**
safe isolation 8, 31–40; notice **38**; preparation 34–8; process 31–4, 38–40
safety 17; connection warning notice 152; and correct functioning 7; equipment 35; requirements 3, 12; tests 138, *see also* health and safety
sampling methodology 49
scheduled inspections 137–8
scheduled test results 137–8
self-certification 14–15
self-employed 13
SELV equipment 153
short answer assessment, on-line 143–5
short circuit current 148
simple circuit voltage drop **133**, 134
single coil clamp test 104
single earth electrode 104
single minor electrical installation 138
single phase circuit 134
single phase installation prospective fault current 126, *126*
single phase (three tests) *39*, 93
skilled person 12–14
skills, practical 12
space access, adequate 29–30
special installations/locations 62, **62**
standard initial test sequence 65–6
standard tests 65, *65*
standardisation 52
statutory information 41–3
steel conduit installation 147
Stroma (UK) 15
sufficient information 44
suitable information 44
suitable tools and clothing 28
supervision 13
supplementary protective bonding 121; system 75–6

supplementary protective bonding
conductors 71, 75–6; continuity **76**
supply, automatic disconnection of 71,
76, 107–12, 121
supply polarity confirmation 100–1

technical experience/knowledge 13–14,
17–18
temperature 88
temporary connections 85, 95
terminals, neutral 99
test equipment 18, 66–9, 152; CAT
IV 100–1; electrical 18, 152; typical
instruments units and range of
values 68, *68*
test leads 36, **36**, 73
test meter 4
test sequence 64–6; standard initial
65–6
test spikes 103–4
testing 63, 66; information 43–4;
polarity 99–101; typical tests for
providing additional protection 123,
123
third-party certification 14–15
three phase circuits 69
three phase installations 20, 93, 129,
148, 153–5; polarity tests 101, *101*;
prospective fault current 126, *127*
three phase process *40*, 93
three-step ring final circuit tests 81
time-served tradesmen 3–4
TN-C-S earthing system: diagram **110**;
identification **110**
TN-C-S supplies 101, 107–10, 152
TN-S earthing system 107–9; diagram
108; identification **109**
tools 28

training 12, 17; options 17–24;
practical 139; process 143
training courses 23–4; college-based
18; formal 137
TT systems **111**; identifying **111**
two pole voltage detectors 36

uninsulated conductors 32–3
United Kingdom Accreditation Service
(UKAS) 67
United Kingdom (UK) 12–13, 107–9,
117; Building Regulations (2010) 42;
City & Guilds 19–23; EAL 19–23;
ELECSA 15; Electricity Safety
Quality and Continuity Regulations
(2002) 42; Health and Safety at
Work Act (1974) 42; LCL 19–23;
NAPIT 15; NICEIC 4, 15; OFQUAL
19–22; Stroma 15
unqualified electricians 99
unreasonable conductors 33
user-selectable controls 131

verification *see* initial verification
visual inspection 55, 99–101, 146
voltage: applicable DC test 92,
92; expected 37; indicator 100;
maximum 37; nominal 148;
nominal supply 134–5; source
39–40
voltage detector 39–40; two pole 36
voltage drop 150; simple circuit **133**,
134; verification 133–5

Wales 14
wiring regulations 9, 18
working precautions 39
written examination 143–5

Printed and bound by CPI Group (UK) Ltd, Croydon, CR0 4YY

21/10/2024

01777110-0001